AF461850

Precision and Accuracy in Biological Crystallography, Diffraction, Scattering, Microscopies, and Spectroscopies

INTERNATIONAL UNION OF CRYSTALLOGRAPHY

BOOK SERIES

IUCr Monographs on Crystallography

1 *Accurate molecular structures*
A. Domenicano and I. Hargittai, editors
2 *P.P. Ewald and his dynamical theory of X-ray diffraction*
D. W. J. Cruickshank, H. J. Juretschke, and N. Kato, editors
3 *Electron diffraction techniques, Vol. 1*
J. M. Cowley, editor
4 *Electron diffraction techniques, Vol. 2*
J. M. Cowley, editor
5 *The Rietveld method*
R. A. Young, editor
6 *Introduction to crystallographic statistics*
U. Shmueli, G. H. Weiss
7 *Crystallographic instrumentation*
L. A. Aslanov, G. V. Fetisov, and J. A. K. Howard
8 *Direct phasing in crystallography*
C. Giacovazzo
9 *The weak hydrogen bond*
G. R. Desiraju and T. Steiner
10 *Defect and microstructure analysis by diffraction*
R. L. Snyder, J. Fiala, and H. J. Bunge
11 *Dynamical theory of X-ray diffraction*
A. Authier
12 *The chemical bond in inorganic chemistry*
I. D. Brown
13 *Structure determination from powder diffraction data*
W. I. F. David, K. Shankland, L. B. McCusker, and Ch. Baerlocher, editors
14 *Polymorphism in molecular crystals*
J. Bernstein
15 *Crystallography of modular materials*
G. Ferraris, E. Makovicky, and S. Merlino
16 *Diffuse X-ray scattering and models of disorder*
T. R. Welberry

17 *Crystallography of the polymethylene chain: An inquiry into the structure of waxes*
D. L. Dorset
18 *Crystalline molecular complexes and compounds: Structure and principles*
F. H. Herbstein
19 *Molecular aggregation: Structure analysis and molecular simulation of crystals and liquids*
A. Gavezzotti
20 *Aperiodic crystals: From modulated phases to quasicrystals*
T. Janssen, G. Chapuis, and M. de Boissieu
21 *Incommensurate crystallography*
S. van Smaalen
22 *Structural crystallography of inorganic oxysalts*
S. V. Krivovichev
23 *The nature of the hydrogen bond: Outline of a comprehensive hydrogen bond theory*
G. Gilli and P. Gilli
24 *Macromolecular crystallization and crystal perfection*
N. E. Chayen, J. R. Helliwell, and E. H. Snell
25 *Neutron protein crystallography: Hydrogen, protons, and hydration in bio-macromolecules*
N. Niimura and A. Podjarny
26 *Intermetallics: Structures, properties, and statistics*
W. Steurer and J. Dshemuchadse
27 *The chemical bond in inorganic chemistry: The bond valence model*, second edition
I. D. Brown
28 *Aperiodic crystals: From modulated phases to quasicrystals: Structure and properties*, second edition
T. Janssen, G. Chapuis, and M. de Boissieu
29 *Biological small angle scattering: Theory and practice*
E. E. Lattman, T. D. Grant, and E. H. Snell
30 *Polymorphism in molecular crystals*, second edition
J. Bernstein
31 *Diffuse X-ray scattering and models of disorder*, second edition
T. R. Welberry

IUCr Texts on Crystallography
1 *The solid state*
A. Guinier and R. Julien
4 *X-ray charge densities and chemical bonding*
P. Coppens
8 *Crystal structure refinement: A crystallographer's guide to SHELXL*
P. Müller, editor
9 *Theories and techniques of crystal structure determination*
U. Shmueli
10 *Advanced structural inorganic chemistry*
W.-K. Li, G.-D. Zhou, and T. Mak
11 *Diffuse scattering and defect structure simulations: A cook book using the program DISCUS*
R. B. Neder and T. Proffen

13 *Crystal structure analysis: Principles and practice*, second edition
W. Clegg, editor
14 *Crystal structure analysis: a primer*, third edition
J. P. Glusker and K. N. Trueblood
15 *Fundamentals of crystallography*, third edition
C. Giacovazzo, editor
16 *Electron crystallography: Electron microscopy and electron diffraction*
X. Zou, S. Hovmöller, and P. Oleynikov
17 *Symmetry in crystallography: Understanding the International Tables*
P. G. Radaelli
19 *Small angle X-ray and neutron scattering from solutions of biological macromolecules*
D. I. Svergun, M. H. J. Koch, P. A. Timmins, and R. P. May
20 *Phasing in crystallography: A modern perspective*
C. Giacovazzo
21 *The basics of crystallography and diffraction*, fourth edition
C. Hammond
22 *Atomic pair distribution function analysis: A primer*
S. J. L. Billinge and K. M. O. Jensen
23 *X-ray structure analysis of bio-macromolecular crystals*
A. Takenaka, editor
24 *Symmetry relationships between crystal structures: Applications of crystallographic group theory in crystal chemistry*, second edition
U. Müller and G. de la Flor
25 *Precision and accuracy in biological crystallography, diffraction, scattering, microscopies, and spectroscopies*
J. R. Helliwell

Precision and Accuracy in Biological Crystallography, Diffraction, Scattering, Microscopies, and Spectroscopies

JOHN R. HELLIWELL

Great Clarendon Street, Oxford, OX2 6DP,
United Kingdom

Oxford University Press is a department of the University of Oxford.
It furthers the University's objective of excellence in research, scholarship,
and education by publishing worldwide. Oxford is a registered trade mark of
Oxford University Press in the UK and in certain other countries.

Published in the United States of America by Oxford University Press
198 Madison Avenue, New York, NY 10016, United States of America

British Library Cataloguing in Publication Data
Data available

Library of Congress Control Number: 2025937492

ISBN 9780198952824

DOI: 10.1093/9780198952848.001.0001

Printed and bound by
CPI Group (UK) Ltd, Croydon, CR0 4YY

Cover image: Prepared by Dr François Jacobs, University of the Free State, Bloemfontein, South Africa

The manufacturer's authorised representative in the EU for product safety is
Oxford University Press España S.A., Parque Empresarial San Fernando de Henares,
Avenida de Castilla, 2 – 28830 Madrid (www.oup.es/en or product.safety@oup.com).
OUP España S.A. also acts as importer into Spain of products made by the manufacturer.

Contents

Preface

Why have I written this book, and in particular as a teaching book?

A common misconception about an experimental method is that it is claimed to be accurate. Such a claim has some justification because every experimentalist strives to eliminate the potential systematic errors in that method. But of course, if we wear the hat of the natural philosopher, as physics was referred to before the 'modern era', we can never determine the truth but only an approximation to it. If, however, we combine one method's results with another method's results we are in a good place because we can assess how well those two sets of results agree. Here we arrive at the domain of accuracy. Where then does that leave the 'single method'? Each method on its own can legitimately be described as precise when the random errors inherent in it have been minimized.

In the teaching of the physics of experimental measurements a common analogy used is that of a dartboard. You can see this if you go to Google Images and search on 'precision and accuracy'. You are inundated with many dartboards as images with each one showing darts in a cluster away from the bullseye centre of the dartboard, thereby being precise but not accurate versus a cluster of poorly placed darts around the centre being imprecise but accurate. It is a striking depiction of precision versus accuracy. However, as a metaphor of the situation of experimental methods the dartboard and its darts gloss over the key fact that in real life we don't know where the bullseye is! The truth is never known.

Where does theory fit into this description? Theory can be a powerful instrument in its own right to discern truth. A classic example is Einstein's posing of the question: what if the speed of light is finite? This question, through his pencil and paper equations, led to consequences and these could be assessed. That theory provides a benchmark, namely the speed of light is indeed finite, and that is the truth. Orthodoxy would require us, however, to state something like: this is a fundamental constant of the universe now but what if its value changed? I have to concede that is an important caveat but nevertheless Einstein's special relativity theory is as close as one can get as an exemplar of truth in physics.

In the modern era a different kind of theory has emerged as we can enjoy the fruits of databases, which in crystallography mean such as the Cambridge Structure Database and the wwPDB (the worldwide Protein Data Bank) and for the latter the highly important amino acid sequences databases complementary to the PDB. A new kind of theoretical analysis has become possible, such as the, by now classic, example of protein three-dimensional (3D) fold prediction. This involves analysing an amino acid sequence against related experimental structures in the PDB, where the precision of each structure is ranked according to the consistent standard of the PDB's Validation Report. Hence phylogenetic changes in a specific sequence suggest conservation of the 3D structure is needed. Thereby a 3D fold for an as yet unknown experimental structure could be derived by prediction algorithms using input from those different databases. The assessment of success, or confidence in the prediction, is the closeness of fit of a prediction to a precise experimental structure where one is available. It is a prediction, though, and thereby a hypothesis of what a 3D structure would look like for an amino acid sequence that has not yet had its structure determined. Experimental data, such as from X-ray crystallography, are needed to know whether it is an appropriate prediction. A major step in crystallography is still always the determination of the crystallography phase of each Bragg reflection. In macromolecular crystallography this was improved enormously, i.e. speeded up, using tuneable synchrotron radiation and the multiple wavelength anomalous dispersion method. Today, however, to use a predicted 3D polypeptide fold is most often a good starting point for phase determination.

The above preamble then provides a context for explaining my motivation for writing this teaching book. It is a book about each experimental method we can use to determine as precisely as possible a 3D structure of a biological macromolecule or portions of it. In this book I also assess how well two or more methods agree in their 3D determined structures. This allows us to dissect our knowledge of the accuracy of what we know. Of course, if we know how accurate our structures are we have a better chance, our best chance, to understand their functions. The dynamics of structures are at the heart of elucidation of function. In the history of our field two competing ideologies existed specific to interaction: the lock and key mechanism of function where rigid entities interacted in 'opening a door' versus the induced fit mechanism as one entity moulded itself into another like a hand into a glove. The latter offers greater structural flexibility. If that is the case, then we must also understand the uncertainties in our determination of atomic positions in our static structures. This is a total void in our PDB entries of atomic coordinates. As data depositors we do not provide either uncertainty estimates on our coordinates or on our atomic B factors. But methods of analysis do exist for each experimental method that we can use to give estimates of these uncertainties. This teaching book will describe these, also assessing their weaknesses as well as strengths. At the very least we must not quote precisions that convey a false confidence in our results; a classic case is the macromolecular crystallography atomic B factor quoted to two decimal places when its precision is only an integer.

I dedicate this book to Durward Cruickshank who profoundly influenced me in our discussions that led up to his seminal paper on the Diffraction Precision Index that he published in 1999 (Cruickshank, D. W. J. (1999), *Remarks about protein structure precision*, Acta Cryst., D55, 583–601). See also Beagley B. and Helliwell J. R. (2018), *Durward W. J. Cruickshank, 7 March 1924–13 July 2007.* Biogr. Mems Fell. R. Soc., 65, 71–87, http://doi.org/10.1098/rsbm.2018.0018.

Acknowledgements

I am grateful to my colleagues on the IUCr OUP Book Series Committee (1996 to 2023) for our many interactions. I am very grateful to the comments I received on my proposal from the current committee membership for this teaching book, and then from the current IUCr Executive Committee who had chance to comment as a second stage, then finally Sonke Adlung and Giulia Lipparini at OUP as a third stage. Throughout this procedure the new Chairman of the Book Series Committee, Dr Jacqui Gulbis, based in Australia, and occasionally on research leave in New York, has been very supportive and helpful to probe my ideas. Thank you, Jacqui, and thank you all.

About the author

John R. Helliwell is a research crystallographer and educator. He was a scientific civil servant supporting the UK and international structural biology and chemistry community principally at the UK's Synchrotron Radiation Source at Daresbury Laboratory (1979–2008), either as a joint appointment or full time. He was also a lecturer at the University of Keele Physics department 1979 to 1983 and at the University of York Physics department 1985 to 1988. From 1989 to 2012 he was a Professor of Structural Chemistry at the University of Manchester where, since 2012, he is an Emeritus Professor of Chemistry. He was awarded a DSc in Physics by the University of York in 1996. His DPhil from Oxford University (1978) was in molecular biophysics specializing in protein crystallography. He has published more than 200 research articles. He has more than 100 data depositions in the PDB, either as collaborator or depositor. In the past decade each PDB deposition has been accompanied by their raw diffraction images placed either at the University of Utrecht or University of Manchester and most recently at Zenodo. He has taught crystallography to undergraduate and graduate physicists, chemists and biochemists at Keele, York, and Manchester Universities as well as to early career researchers in Europe and the USA at crystallography society training courses. He was a member of the IUCr OUP Book Series Committee from 1996 to 2023, including as its Chairman from 2017 to 2023. He has published a research monograph within the Book Series as well (*Macromolecular Crystallization and Crystal Perfection* with Naomi Chayen and Edward Snell in 2010).

1
Introduction

"*All models are wrong, but some models are useful*", as neatly stated by Box (1976). Statistical models always fall short of the complexities of reality but can still be useful nonetheless. The aphorism originally referred just to statistical models, but it is now sometimes used for scientific models in general. Whilst this may suit statisticians, in the experimental sciences this translates into the need, always, to estimate the precision and accuracy of our model derived from experiment. But is 'always' adhered to in structural biology? Even worse than not always is 'never' in terms of the wwPDB entries. Yet the PDB is an incredibly useful compilation of current knowledge of experimentally derived structures from biological crystallography (mainly X-ray crystallography but also growing quickly are microED as well as neutron crystallography for protonation details of structures), NMR spectroscopy, and electron cryo-microscopy (cryoEM). These were recently supplemented by the predictions of 3D structures from gene sequences by AlphaFold2, each accompanied by a confidence prediction of their quality.

The absence of error estimates on an atom's coordinates in a protein crystal structure was highlighted by Cruickshank (1999). His thinking was extended to include B factors (Helliwell (2023)) and his coordinate error estimates are widely available via a webtool set up in Bangalore at the Molecular Biophysics and Bioinformatics research centres (Kumar et al (2015)). These approaches can be extended to all the structural biology methods. After all what is the scientific truth in this field without error estimates?

Structural biology's foundation is chemical crystallography launched in 1913 by W L Bragg with the first X-ray crystal structures, namely of the alkali halides, notably sodium chloride (Bragg (1913)). A key role in this elucidation was the importance of the mass measurement.

Mass spectrometry has expanded enormously in its precision and its applications in analytical chemistry are now routine (Ebsworth et al (1991)). Its role in structural biology has also developed strongly (Liko et al (2016)). The

Precision and Accuracy in Biological Crystallography, Diffraction, Scattering, Microscopies, and Spectroscopies. John R. Helliwell, Oxford University Press. © John R. Helliwell (2025). DOI: 10.1093/9780198952848.003.0001

mass is measured prior to crystallization, but the crystallization conditions add uncertainties to quite what has been crystallized (Chayen et al (2010)).

A key feature of biological structure determination involves a range of validation methods. These are linked with chemical crystallography methods but have had to be expanded to bring in a wider range of information sources. A notable step was the recognition of the Ramachandran plot of the dihedral angles distribution of a protein's polypeptide backbone (specifically the phi and psi angles) to be expected of a protein's polypeptide chain. Many more checks have been added to the Protein Data Bank's Validation Report (Kleywegt (2012)).

Another validation tool is the so-called ***round-robin project.*** These have been widely applied over the decades and are very successful in showing up just what precision and accuracy can be achieved in structure determination. One very recent example for solution X-ray scattering was reported by Trewhella et al (2022).

The probes available to crystallography, diffraction, and scattering are of course X-rays, the first to be applied, then electrons, and finally neutrons. The latter was slowest to develop as the available sources and their beam intensities were too low to be useful. However, as the neutron source instrumentation, methods, and software have been expanded, neutron crystallography results have appeared more and more and are regarded as a reference benchmark, being free of irradiation damage effects on a sample (Helliwell (2021)).

As we move away in this book's treatment to other states of matter the assessments of precision and accuracy must adapt to the circumstances. So, specialized validation approaches have grown up within fibre diffraction, powder diffraction, solution scattering (and which vary within the X-ray and neutrons domains), electron microscopy of single particles, X-ray absorption spectroscopy of all states of matter, and NMR spectroscopy of solutions and in the solid state. A comprehensive treatment of the various biological structure techniques are the chapters in International Tables Volume F, 2nd edition (Arnold et al (2012)). The most up-to-date cryoEM validation checks applied by the PDB are described by Kleywegt et al (2024). IUCr Journal *Acta Cryst F* at its launch worked closely with leading NMR spectroscopists to help define validation criteria for NMR macromolecular structures (Einspahr and Guss (2008)).

In traditional physics measurements it is a standard approach that combining methods allows for accuracy assessments and this is described in detail in Chapter 15, each method being precise within its own analyses. Where feasible, combining methods is done in the structural biology field more and more. Structural chemistry as a field adopted this practice many years before (Ebsworth et al (1991)).

Predictive methods are of course vital not only as an approach to plan new experiments but to fill in gaps in accessible experimental knowledge. How does one assess the confidence of these? This assessment is a major challenge, yet the AlphaFOLD2 'protein folding prediction' team describe their predictions as accurate (Jumper et al (2021)). When queried it was explained to me that this was because they had a confidence estimate, which is very good but not, however, a criterion of accuracy as per the physicists' criterion. Another domain of prediction in structural biology is the very targeted theme of prediction attempts of the protonation states of ionizable amino acids. The predictions have been evaluated by direct comparison with the experimental methods of neutron protein crystallography, ultra-high resolution X-ray protein crystallography, and NMR titration studies (Fisher et al (2009)).

Experimental and theoretical/computational methods are combined, not only to assess accuracy, but also to assemble, like a jigsaw puzzle, a model of complex systems such as an organelle in a cell or indeed the whole cell (see Ward et al (2013) for an early impetus on this approach). These 'completed jigsaw puzzles' can be guided by the relatively new experimental method of cryo-electron tomography where a large portion of a whole cell is directly imaged (Andrews et al (2022)).

In seeking to understand biological processes, dynamics simulations start from a static structure to explore timescales, and complement experimental structural dynamics studies such as using flow cells or photocrystallography. The molecular dynamics approach works with Newton's law of motion $F=ma$ and where atomic movements are explored according to a force field. An alternative is to assemble animated cartoons based on a sequence of individual static structures. Here, however, in what order does one place the frames of the cartoon? A simple and reasonable assumption is that the time-series will follow a criterion of the smallest incremental structure changes step by step. It is obvious that an average of these changes frame to frame will not necessarily allow for large movements of loops of a macromolecule to be confidently visualized. Basically the 'frames' of a cartoon can miss a lot of molecular motion. Nevertheless, the cartoon is the one computational approach that is based on the most precise individual experimental structures such as from crystallography, cryoEM, or NMR.

Overall, we work within the criticisms of holistic biologists that molecular biology methods are reductionist.

Key Learning Point

- This overview introductory chapter sets the scene for the chapters to come in this book.

References

Andrews, B., Chang, J. B., Collinson, L., et al (2022). *Imaging cell biology*. Nat Cell Biol, 24, 1180–1185. https://doi.org/10.1038/s41556-022-00960-6

Arnold, E., Himmel, D. M., and Rossmann, M. G. (Editors) (2012). *International Tables for Crystallography, Volume F: Crystallography of Biological Macromolecules*, 2nd Edition. Part 19 *Other Experimental Techniques*. Wiley, Chichester, pp. 553–632. ISBN: 978-0-470-66078-2.

Box, G. E. P. (1976). *Science and statistics*. Journal of the American Statistical Association, 71 (356), 791–799. doi:10.1080/01621459.1976.10480949

Bragg, W. L. (1913). *The structure of some crystals as indicated by their diffraction of X-rays*. Proc. R. Soc. London Ser. A, 89, 248–277.

Chayen, N. C., Helliwell, J. R., and Snell, E. H. (2010). *Macromolecular Crystallization and Crystal Perfection* (International Union of Crystallography 'Monographs on Crystallography'). Oxford University Press, Oxford. ISBN-9780199213252 doi:10.1093/acprof:oso/9780199213252.001.0001

Cruickshank, D. W. J. (1999). *Remarks about protein structure precision*. Acta Cryst., D55, 583–601.

Ebsworth, E. A. V., Rankin, D. W. H., and Cradock, S. (1991). *Structural Methods in Inorganic Chemistry*, 2nd edition. Blackwell, Oxford. 528 pages.

Einspahr, H. and Guss, M. (2008). *Validation of macromolecular structures: Updating standards for publication of NMR structures in an IUCr journal*. Acta Cryst., F64, 63. https://journals.iucr.org/services/nmr/nmrmeetingsummary.html

Fisher, S. J., Wilkinson, J., Henchman, R. H., and Helliwell, J. R. (2009). *An evaluation review of the prediction of protonation states in proteins versus crystallographic experiment*. Crystallography Reviews, 15(4), 231–259. https://doi.org/10.1080/08893110903213700.

Helliwell, J. R. (2021). *Combining X-rays, neutrons and electrons, and NMR, for precision and accuracy in structure–function studies*. Acta Cryst., A77, 173–185.

Helliwell, J. R. (2023). *Error estimates in atom coordinates and B factors in macromolecular crystallography*. Current Research in Structural Biology, 6, 100111. ISSN 2665-928X. https://doi.org/10.1016/j.crstbi.2023.100111.

Jumper, J., Evans, R., Pritzel, A., Green, T., Figurnov, M., Ronneberger, O., Tunyasuvunakool, K., Bates, R., Žídek, A., Potapenko, A., Bridgland, A., Meyer, C., Kohl, S. A. A., Ballard, A. J., Cowie, A., Romera-Paredes, B., Nikolov, S., Jain, R., Adler, J., Back, T., Petersen, S., Reiman, D., Clancy, E., Zielinski, M., Steinegger, M., Pacholska, M., Berghammer, T., Bodenstein, S., Silver, D., Vinyals, O., Senior, A. W., Kavukcuoglu, K., Kohli, P., and Hassabis, D. (2021). *Highly accurate protein structure prediction with AlphaFold*. Nature, 596, 583–589.

Kleywegt, G. J. (2012). *Validation of Protein Crystal Structures.* Chapter 21.1 in *International Tables for Crystallography, Volume F: Crystallography of Biological Macromolecules,* 2nd Edition. Arnold, E., Himmel, D. M., and Rossmann, M. G. (Editors). Wiley, Chichester.

Kleywegt, G. J., Adams, P. D., Butcher, S. J., Lawson, C. L., Rohou, A., Rosenthal, P. B., Subramaniam, S., Topf, M., Abbott, S., Baldwin, P. R., Berrisford, J. M., Bricogne, G., Choudhary, P., Croll, T. I., Danev, R., Ganesan, S. J., Grant, T., Gutmanas, A., Henderson, R., Heymann, J. B., Huiskonen, J. T., Istrate, A., Kato, T., Lander, G. C., Lok, S.-M., Ludtke, S. J., Murshudov, G. N., Pye, R., Pintilie, G. D., Richardson, J. S., Sachse, C., Salih, O., Scheres, S. H. W., Schroeder, G. F., Sorzano, C. O. S., Stagg, S. M., Wang, Z., Warshamanage, R., Westbrook, J. D., Winn, M. D., Young, J. Y., Burley, S. K., Hoch, J. C., Kurisu, G., Morris, K., Patwardhan, A., and Velankar, S. (2024). *Community recommendations on cryoEM data archiving and validation.* IUCrJ, 11, 140–151.

Kumar, K. S. D., Gurusaran, M., Satheesh, S. N., Radha, P., Pavithra, S., Thulaa Tharshan, K. P. S., Helliwell, J. R., and Sekar, K. (2015). *Online_DPI: A web server to calculate the diffraction precision index for a protein structure.* J. Appl. Cryst., 48, 939–942.

Liko, I., Allison, T. M., Hopper, J. T. S., and Robinson, C. V. (2016). *Mass spectrometry guided structural biology.* Curr. Opin. Struct. Biol., 40, 136–144.

Trewhella, J., Vachette, P., Bierma, J., Blanchet, C., Brookes, E., Chakravarthy, S., Chatzimagas, L., Cleveland, T. E., Cowieson, N., Crossett, B., Duff, A. P., Franke, D., Gabel, F., Gillilan, R. E., Graewert, M., Grishaev, A., Guss, J. M., Hammel, M., Hopkins, J., Huang, Q., Hub, J. S., Hura, G. L., Irving, T. C., Jeffries, C. M., Jeong, C., Kirby, N., Krueger, S., Martel, A., Matsui, T., Li, N., Perez, J., Porcar, L., Prange, T., Rajkovic, I., Rocco, M., Rosenberg, D. J., Ryan, T. M., Seifert, S., Sekiguchi, H., Svergun, D., Teixeira, S., Thureau, A., Weiss, T. M., Whitten, A. E., Wood, K., and Zuo, X. (2022). *A round-robin approach provides a detailed assessment of biomolecular small-angle scattering data reproducibility and yields consensus curves for benchmarking.* Acta Cryst., D78, 1315–1336.

Ward, A. B., Sali, A., and Wilson, I. A. (2013). Integrative structural biology. Science, 339, 913–915. doi:10.1126/science.1228565.

2
The physics of errors as illustrated by X-ray crystallography

Experimental science endeavours to gain insight into physical reality through measurements and then interpretation of those measurements using models. Physical reality is in itself an interesting concept as it implies that there is a truth. Let's set these general remarks in the context of X-ray crystallography as an example before the ensuing chapters look at a wide variety of other methods determining the structure of biological macromolecules.

2.1 Random and systematic errors in X-ray crystallography

Some of the systematic effects listed in Table 2.1 can be minimized and some, such as X-ray radiation damage, cannot. The inability to eliminate systematic errors means that any measurement is going to be inaccurate. For the random errors we can seek the best precision possible by having as many counts as possible in the measured intensity of each reflection.

2.2 The role of expertise and reproducibility

A way of testing the level of expertise in X-ray crystallography measurement is to have individual laboratories measure from the same crystal sample, although radiation damage becomes a factor if one crystal is used by the laboratories involved, or to study different crystals from the same batch identically prepared. This is the round-robin type of study and brings in not only the different diffractometers of each laboratory, but also if there are any human factors involved they can be determined. This type of study across various methods is described in Chapter 6.

A further way of testing the reproducibility of the final atomic model of the crystal unit cell is to use different software at each step, presuming that there are such available. So, the processing of the raw diffraction images to the set of F_{meas} is done and then use is made of each set of F_{meas} to refine the atomic

Precision and Accuracy in Biological Crystallography, Diffraction, Scattering, Microscopies, and Spectroscopies. John R. Helliwell, Oxford University Press. © John R. Helliwell (2025). DOI: 10.1093/9780198952848.003.0002

Table 2.1 This table provides an extensive list of random and systematic errors in X-ray crystallography measurements of the reflection intensities and their associated structure factor amplitudes, F_{meas}.

Random	Systematic
Poisson distribution in the arrival of quanta in the incident X-ray beam.	Differences between measured and actual crystal dimensions.
Short-term variations in incident X-ray beam flux.	Incorrect transmission coefficient.
Short-term variations in sensitivity of detection system.	Incorrect extinction coefficient.
	Non-monochromatic incident X-ray beam.
	Thermal diffuse scattering.
	Multiple scattering.
	Inhomogeneous crystal specimen.
	Radiation damage.
	Mechanical misalignments.
	Long-term variation in incident beam flux.
	Long-term variation in sensitivity of detection system.

From Abrahams (1969) with the permission of the IUCr Journals.

model again with different software against each set of those F_{meas}. The distinct models produced can then be compared. Figure 2.1 shows such multiple pathways of data analysis through to model.

2.3 Prior knowledge harnessed as model refinement restraints

All the covalent bond distances and angles that form a biological macromolecule are available from high-resolution chemical crystal structures and so we can use this prior knowledge as restraints in a new macromolecular refinement. In biological crystallography this is a major advantage because the available number of F_{meas} may barely exceed the number of parameters to be determined. Refinement methods involve the determination of shifts to the atomic parameters (coordinates and atomic displacement parameters, 'B factors') to agree better with the observed diffraction data whilst preserving the known stereochemical features of proteins and nucleic acids. This is achieved by minimizing a composite observational function:

$$\Phi = \Phi_{X\text{-}ray\ diffraction} + \Phi_{bonds} + \Phi_{bond\ angles} + \Phi_{torsion\ angles} + \Phi_{non\text{-}bonded\ interactions}$$

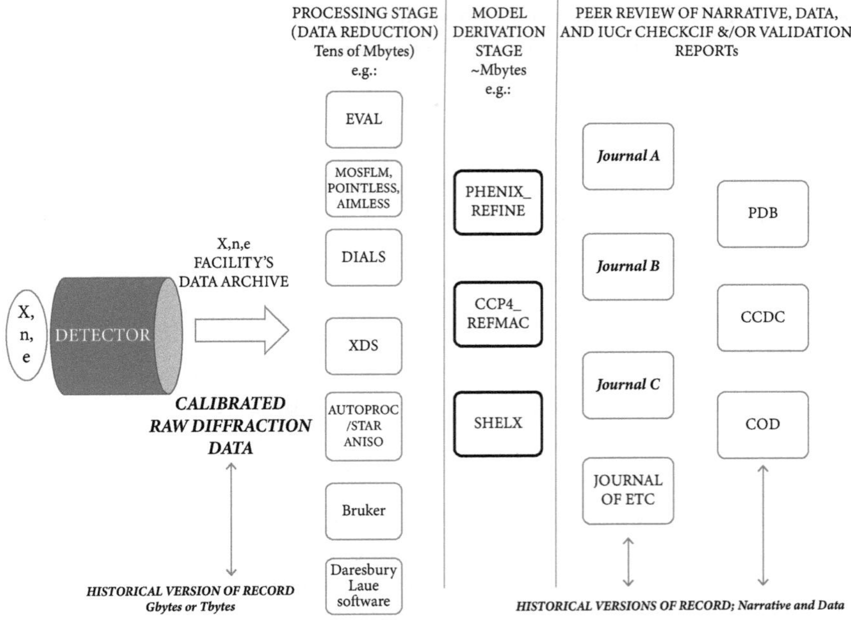

Figure 2.1 The workflow possibilities in single crystal raw diffraction image data processing and molecular model refinement, moving from left to right. Alternative choices of software are labelled. At far right are the options of the journal possibilities and appropriate deposition databases for coordinates and data. Note that there is also EMDB for cryoEM raw images (see https://www.ebi.ac.uk/emdb/), thereby equivalent to raw diffraction images archiving.

where

$$\Phi_{\text{X-ray diffraction}} = \sum_{hkl} \frac{1}{\sigma_F^2}(F_{\text{obs}} - F_{\text{calc}})^2$$

$$\Phi_{\text{bonds}} = \sum_{\text{distances}} \frac{1}{\sigma_D^2}(d_{\text{ideal}} - d_{\text{model}})^2$$

$$\Phi_{\text{bond angles}} = \sum \frac{1}{\sigma^2}(\tau_{\text{ideal}} - \tau_{\text{model}})^2$$

$$\Phi_{\text{torsion angles}} = \sum \frac{1}{\sigma^2}\left[1 + \cos(m\,\theta_{\text{i}} + \delta)\right]$$

$$\Phi_{\text{non-bonded interactions}} = \sum \left(\frac{A}{r_i^{12}} + \frac{B}{r_i^{6}}\right)$$

The terms, besides the X-ray diffraction term, are collectively known as the energy terms. The relative weight of the X-ray and the energy terms can be varied. The number of parameters to be refined is large. Because of the limited resolution of the X-ray diffraction pattern of a protein crystal the energy terms are exceedingly useful to increase the data-to-parameter ratio.

2.4 Standard uncertainty estimates

A very helpful paper describing the uncertainties of measurement as well as prior molecular knowledge and a wide variety of definitions is that by Schwarzenbach et al (1989). It is a report of the International Union of Crystallography's Subcommittee on Statistical Descriptors and thereby also offers recommendations on good practice within an overarching guideline that:

> ***Thoughtless use of established procedures in widely distributed software may be as harmful as the natural tendency of most people to prefer results in agreement with preconceived ideas.*** *Note, however, that preconceived ideas are an ingredient of Bayesian statistics. Since the precision may be evaluated with greater confidence than the accuracy, it is not surprising that the results of independent determinations of the same structure may differ by much more, and hardly ever by less, than is allowed by statistical tests.*

The words in bold above are those of the original publication. This text is reproduced with the permission of IUCr Journals. An update of this report was published by Schwarzenbach et al (1995) which recommended '*replacement of the term estimated standard deviation (e.s.d.) by standard uncertainty (s.u.) or by combined standard uncertainty (c.s.u.) in statements of the statistical uncertainties of data and results*' and requested '*a complete description of the experimental and computational procedures used to obtain all results submitted to IUCr publications*'. This text quote is also reproduced with the permission of IUCr Journals.

A major limitation of a protein structure model is the general absence of error estimates on coordinates or B factors (these latter are more formally referred to as atomic displacement parameters). Reliable $\sigma(x)$ values are needed for any discussion of non-dictionary distances between atoms in different residues, between protein and solvent atoms, or between metal atoms and their ligands. Because of this Cruickshank (1999) developed a formalism which he called a Diffraction Precision Index (DPI). It gives an overall precision of a protein structure. The precision estimate of an average atom coordinate error in a biological macromolecule crystal structure, with an average B factor i.e. B_{avg} is:

$$\sigma\left(x, B_{avg}\right) = 1.0(N_i/p)^{1/2}C^{1/3}Rd_{min} \qquad \text{(Eq.1)}$$

Here, $p = \left(n_{obs} - n_{params}\right)$, R is the usual residual $\sum|\Delta F| / \sum|F|$ and N_i is the number of atoms of type i. C is the diffraction data completeness, now increasingly very close to 1.0. d_{min} is the diffraction resolution quoted by the depositing authors for the respective study under question at the PDB. This parameter d_{min} is subject to rather arbitrary values, but which has been put on a firm footing by Diederichs and Karplus (2013) who evaluated when the R_{free} (Brunger (1992)) starts to deteriorate upon adding ever higher resolution (i.e. weaker) diffraction data. Another important role of R_{free} in the current context, instead of R, allowed replacing $\left(n_{obs} - n_{params}\right)$ with n_{obs} alone (Cruickshank (1999), in sections 6.3 and 7.3) is when the number of observations dips below the number of model refinement parameters. This situation could occur at the lower diffraction resolutions (~3 Å or worse).

Gurusaran et al (2014), encouraged by the implications of Cruickshank (1999), made the necessary extension to individual atoms to calculate the precision found on each atomic coordinate for biological macromolecules, taking into account the B factor of an individual atom versus that of an average atom:

$$\text{Coordinate error of an atom} = \text{DPI}\left(B_{atom}/B_{average}\right)^{1/2}$$

Additionally, Kumar et al (2015) introduced a webserver that provides a transformed PDB entry with individual atom coordinate errors derived from applying the DPI method using the parameters provided by the authors. This webserver has been extensively used and harnessed in describing non-covalent distance error estimates as well as assessing the significance, or otherwise, of atom movements in a variety of studies.

The question arises: does the addition of covalent bond distances and angles to the whole model refinement as restraints improve the precision of determining positions of the non-covalently bonded atoms? Cruickshank (1999) addressed this in several points as follows:

- *The application of restraints in protein refinement does not affect the key idea about the method of error estimation.*

- *Low-resolution structures can validly be determined by using restraints, even though the number of diffraction observations is fewer than the number of atomic coordinates. Here the number of parameters may exceed the number of diffraction data. This difficulty can be circumvented empirically by replacing p with n_{obs} and R with R_{free}.*
- *Compared with unrestrained refinements for small molecules, in restrained refinements for proteins there is a major numerical distinction between the standard uncertainty (s.u.) $\sigma(x_i)$ of an atomic coordinate and the s.u.$\sigma(l_{ij})$ of a bond length. The atomic $\sigma(x_i)$ depends strongly on B.*
- *Small root mean square (rms) differences between refined and dictionary bond lengths are not an indication of $\sigma(x)$ quality.*

Blow (2002) reformulated the Cruickshank (1999) treatment into experimental parameters and commented on restraints as follows:

- *For protein crystallography at low resolution, refinement is carried out with restraints and successful refinement is possible even if p is negative. For this situation, Cruickshank (1999) 'empirically' proposes the use of R_{free} (Brunger, 1992) in place of R and n_{obs} in place of p.*
- *refinement of individual atomic positions and B factors without restraints is impossible for a typical protein if the resolution is worse than about 2.6 Å.* Cruickshank (1999) *notes cases where p is negative for data at 2.6 and 2.5 Å resolution, respectively. (Standard practice requires a considerable excess of measurements over variables, say three times as many, and unrestrained refinement is inadvisable if d_{min} is worse than about 1.8 Å for a typical protein.)*

The above text quotations from Cruickshank (1999) and Blow (2002) are with the permission of IUCr Journals.

The standard uncertainties on a biological macromolecule's atomic displacement parameters (the 'B factors') has been an entirely different challenge but is obviously important since the crystallographic community has developed the habit of quoting B factors to a false precision in papers. This can convey a false certainty in the dynamics of a structure. A method involving parallelization of workflows for diffraction image data processing does, however, offer estimates of the precision of B factors (see Figure 2.2). Furthermore, this approach allows a further check on the Cruickshank (1999) coordinate error estimates, as well as the precision of B factors. These matters are elaborated on by Helliwell (2023).

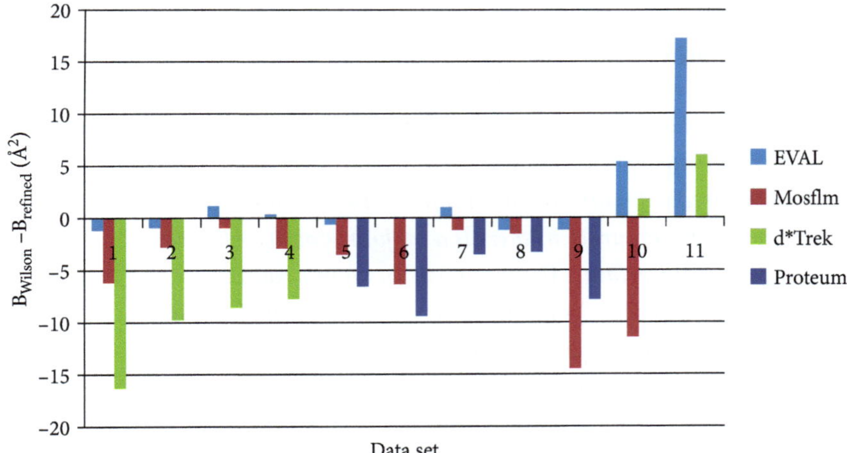

Figure 2.2 A plot showing the difference between experimental B factors and those calculated from the molecular model ($B_{\text{Wilson}} - B_{\text{refined}}$) (Å^2) for each of 11 protein (hen egg white lysozyme) crystals, using four different diffraction image processing packages. Tanley et al (2013).
Reproduced with the permission of Dr Loes Kroon-Batenburg and the IUCr Journals.

There are limitations in applying these estimates to macromolecular crystal structures from other laboratories: (i) Checks must be made that gross errors of atom misidentification (see Chapter 7), or inclusion or exclusion when unwarranted by the evidence, are not present. Likewise, an incorrect choice of resolution limit should have been avoided. If the raw diffraction data are made available as well, one can check the estimated resolution limit with the Diederichs and Karplus (2013) method. (ii) Comparisons between crystal structures at widely different temperatures, such as 100 K to room temperature, are difficult to apply because of the variation in atomic displacement parameters between these temperatures. In addition, there are changes in the structure and dynamics on cooling, which we have noted (see Deacon et al (1997)) as well as others (explained from a physical chemistry viewpoint by Halle (2004) and reviewed recently by Fischer (2021)). (iii) An isotropic B factor is not an ideal physics-based descriptor (see e.g. Ploscariu et al (2021)) as it includes both temporal and spatial variations in its single atomic displacement parameter estimate. Disorder situations were specifically excluded from Cruickshank's (1999) DPI analysis. Within those situations split occupancy order (often referred to as 'static disorder' in chemical crystallography) should, however, be treatable by the DPI approach.

2.5 How can X-ray crystallography make its results as close to truth as possible?

As stressed earlier, a single method such as X-ray crystallography can be as precise as possible by reducing the impact of random errors. But can it also generate results as close to the truth as possible or, more simply put, optimize its likely agreement with other experimental methods? To do this it must reduce as much as possible the impact of systematic errors. Table 2.1's list of systematic errors cites radiation damage by the X-ray beam itself. In biological X-ray crystallography this has most usually been moderated by cryo-cooling the crystal sample to 100 K. This greatly reduces secondary radiation damage of electron free radicals generated by the absorbed X-rays if not the primary damage arising from the absorption of the X-rays itself. However, it introduces a new systematic error; that of being far from the working temperature of the biological macromolecule in the living organism. The nature of this new systematic error is that structural artefacts at 100K occur (Deacon et al 1997, Halle 2004). Ways must be found, ideally, to determine the structures of these macromolecules at their living organism's working temperature, which for humans means 37 °C (Jacobs et al 2024).

Key Learning Points

- Measurement physics can treat errors, both random and systematic. The literature on X-ray crystallography provides a useful background, and key references are cited.
- Extensions in macromolecular crystallography for atomic coordinate uncertainties and B factor uncertainties are made.
- Parallel workflows based on the same set of raw diffraction images can be used to compare the atomic models and thereby make estimates of the errors on the B factors as well as make a cross check on the Cruickshank (1999) coordinate error estimates.

A further systematic error of using X-rays is in the nature of this probe itself, namely that it is rather insensitive to hydrogens. To determine the positions of protons, X-rays have to be complemented by the use of another probe, most often neutrons. Even this latter probe requires that hydrogens in the biological macromolecule be changed to deuteriums, so as to harness the positive signature of deuteriums for neutrons versus the negative one of hydrogen. There is little to no disadvantage (effect) of this H to D swap on structure (Fisher and Helliwell (2008)), but any interest in their kinetics will have to allow for the slower kinetics of deuterium.

References

Abrahams, S. C. (1969). *Indicators of accuracy in structure factor measurement.* Acta Cryst., A25, 165–173.

Blow, D. M. (2002). *Rearrangement of Cruickshank's formulae for the diffraction-component precision index.* Acta Cryst., D58, 792–797.

Brunger, A. (1992). *Free R value: a novel statistical quantity for assessing the accuracy of crystal structures.* Nature, 355, 472–475.

Cruickshank, D. W. J. (1999). *Remarks about protein structure precision.* Acta Cryst., D55, 583–601.

Deacon, A., Gleichmann, T., Kalb (Gilboa), A. J., Price, H., Raftery, J., Bradbrook, G., Yariv, J., and Helliwell, J. R. (1997). *The structure of concanavalin A and its bound solvent determined with small-molecule accuracy at 0.94Å resolution.* Faraday Transactions, 93 (24), 4305–4312.

Diederichs, K. and Karplus, P. A. (2013) *Better models by discarding data?* Acta Cryst., D69, 1215–1222.

Fischer, M. (2021). *Macromolecular room temperature crystallography.* Q. Rev. Biophys, 54, 1. https://doi.org/10.1017/S0033583520000128.

Fisher, S. J. and Helliwell, J. R. (2008). *An investigation into structural changes due to deuteration.* Acta Cryst., A64, 359–367.

Gurusaran, M., Shankar, M., Nagarajan, R., Helliwell, J. R., and Sekar, K. (2014). *Do we see what we should see? Describing non-covalent interactions in protein structures including precision.* IUCrJ, 1, 74–81.

Halle, B. (2004). *Biomolecular cryocrystallography: Structural changes during flash-cooling.* Proc. Natl. Acad. Sci. U.S.A., 101, 4793–4798.

Helliwell, J. R. (2023). *Error estimates on atom coordinates and B factors in macromolecular crystallography.* Current Research in Structural Biology, 100111, ISSN 2665-928X, https://doi.org/10.1016/j.crstbi.2023.100111.

Jacobs, F. J. F., Helliwell, J. R., and Brink, A. (2024). *Body temperature protein X-ray crystallography at 37°C: A rhenium protein complex seeking a physiological condition structure.* Chemical Communications, *60*(95), 14030–14033. doi: 10.1039/D4CC04245J.

Kumar, K. S. D., Gurusaran, M., Satheesh, S. N., Radha, P., Pavithra, S., Thulaa Tharshan, K. P. S., Helliwell, J. R., and Sekar, K. (2015). *Online_DPI: A web server to calculate the diffraction precision index for a protein structure.* J. Appl. Cryst., 48, 939–942.

Ploscariu, N., Burnley, T., Gros, P., and Pearce, N.M. (2021). *Improving sampling of crystallographic disorder in ensemble refinement.* Acta Cryst., D77, 1357–1364.

Schwarzenbach, D., Abrahams, S. C., Flack, H. D., Gonschorek, W., Hahn, Th., Huml, K., Marsh, R. E., Prince, E., Robertson, B. E., Rollett, J. S., and Wilson, A. J. C. (1989). *Statistical descriptors in crystallography: Report of the IUCr Subcommittee on Statistical Descriptors.* Acta Cryst., A45, 63–75.

Schwarzenbach, D., Abrahams, S. C., Flack, H. D., Prince, E., and Wilson, A. J. C. (1995). *Statistical descriptions in crystallography. II. Report of a Working Group on Expression of Uncertainty in Measurement.* Acta Cryst., A51, 565–569.

Tanley, S. W. M., Schreurs, A. M. M., Helliwell, J. R., and Kroon-Batenburg, L. M. J. (2013). *Experience with exchange and archiving of raw data: comparison of data from two diffractometers and four software packages on a series of lysozyme crystals.* J. Appl. Cryst., 46, 108–119.

3

History of the reliability of structure determination methods

In the history of crystal structure analysis, a major methodological transition was the introduction by Hughes (1941) of least squares model refinement against diffraction data. As Jim Ibers remarked in his American Crystallographic Association (ACA) Biographical sketch '*Edward Hughes in the Pauling Group in 1941 was the first to apply the least-squares technique to the refinement of crystal structures*'. Hughes (1941) used an '*International Business Machines Co. Tabulator using the Hollerith punched card system*' instead of the manual Beevers Lipson strips that he reported in his publication the year before (Hughes (1940)). It was in Hughes (1940) though that he mentioned the word 'reliability'. This was with respect to the measured intensities rather than the molecular model. Whilst Hughes (1940, 1941) tabulates the measured structure factor amplitudes, and the corresponding values calculated from the molecular model, an overall residual was not calculated. Hughes (1941) emphasized the practical details of the calculation (see Figure 3.1), for the melamine nine non-hydrogen atoms crystal structure, as follows: '*The cards were punched, verified, and the normal equations produced in slightly less than two days. The resulting normal equations consisting of eighteen simultaneous equations in the eighteen parameters were solved by an iteration method in about four hours*'. Interest in molecular model refinement was evidently stirred by the Hughes (1941) paper and other variants followed. Cruickshank (1952) compared the relationship between the Fourier and least squares methods. The Fourier method was developed by Booth (1945, 1946, 1947). The method of Booth involved corrections to atomic parameters in real space based on a difference Fourier map.

To come back to the word 'reliability', A J C Wilson's (1950) article focused on the molecular model and opened with '*The reliability index* $R \equiv (\sum \| F_{obs.} | - | F_{calc.} \|) / (\sum |F_{obs.}|)$ *is widely used as a test of the quality of a structure determination*'. The reliability index can be more simply called the least squares residual, which is not then judgemental. Cruickshank

Precision and Accuracy in Biological Crystallography, Diffraction, Scattering, Microscopies, and Spectroscopies. John R. Helliwell, Oxford University Press. © John R. Helliwell (2025). DOI: 10.1093/9780198952848.003.0003

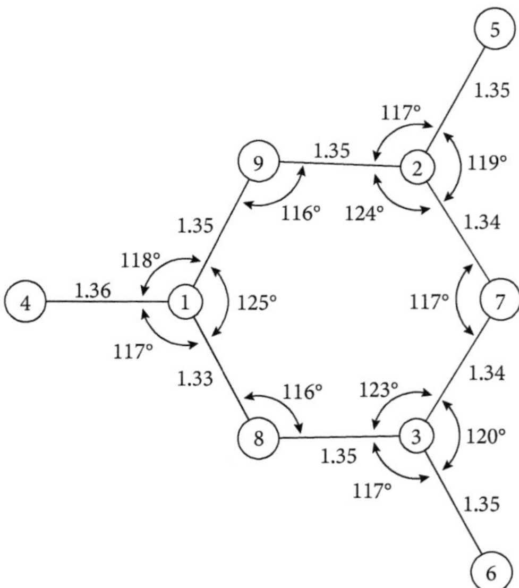

The melamine molecule: small circles carbons, large circles nitrogens, hydrogens not shown. The molecule is coplanar. Bond lengths are in Ångström units.

Figure 3.1 This figure of Hughes (1941) is shown with its original caption. On the precision of these bond distances shown in this figure, Hughes (1941) remarks in his main text that '*The displacements of the atoms from the average plane of the molecule are all within the limits of the probable error, which is estimated to be about* +/- 0.015 Å *for the position of an atom and about* +/– 0.02 Å *for the length of a bond. It is thus not very likely that any bond length is in error by more than* 0.05 Å'.

Reprinted (adapted) with permission from Hughes, E. W. (1941). J. Am. Chem. Soc., 63, 1737–1752.

(1960) discussed the requirements necessary for (i) determining bond lengths crystallographically within a limit of error of 0.01Å, (ii) the required precision for X-ray diffraction intensities, and (iii) gave a simple approximate formula relating the residual R to the coordinate estimated standard deviation. Whilst reliability is implicit in the considerations discussed it is not explicitly used, in contrast to Wilson (1950).

The procedure for a crystal structure analysis that is generally used today, of course, involves first of all a solution to the phase problem, then a difference Fourier electron density map to locate any missing atoms, or indicate disordered moieties, and finally a molecular model refinement as well as the addressing of checkCIF or PDB Validation report alerts by the crystallographer (see e.g. the book by Giacovazzo et al (2011)).

Hughes (1941) concluded as follows:

> *Summary*
>
> *The crystal structure of melamine has been investigated. The monoclinic unit cell has* a = *10.54 Å,* b = *7.45 Å,* c = *7.25 Å, ß* = 112°2′. *The space group is $P2_1/a$, and there are four molecules per cell. A new method for refining parameters, based upon least squares, has been described and was used in conjunction with Fourier syntheses to locate the atoms.*

He also has an acknowledgement: '*In conclusion I must thank Professor (Linus) Pauling*'.

Brief biographical sketch: E. W. Hughes (1924–1979) was based at Caltech in Pasadena, California, USA when he published his 1940 and 1941 papers in JACS. He served as ACA President in 1954. An extensive description of the academic career and publications of E. W. Hughes can be found here: http://pdf.oac.cdlib.org/pdf/caltech/hughese.pdf.

The model refinement of biological macromolecules presented different challenges. A summary article (Murshudov et al (1997)), which links to those early calculation methods, states: '*It was recognised in the 1960s that macromolecular refinement posed special problems. There were too few observations to refine the atomic parameters using least squares minimization alone, and the calculation of the structure factors and derivatives from such a large number of coordinates challenged the computing resources available*.' A key help, to add observations besides the diffraction data, was the availability of dictionary values of bond distances and angles from chemical crystallography that could act as restraints (Konnert (1976), Konnert and Hendrickson (1980)). Murshudov et al (1997) considered the reliability of coordinates within a maximum likelihood formalism for the refinement. The assumption that different parts of a structure might have different errors was considered. Cruickshank (1999) introduced the Diffraction Precision Index to provide an overall coordinates precision of a protein crystal structure based on the processed X-ray diffraction data. This was extended to non-bonded individual atoms by Gurusaran et al (2014) and Kumar et al (2015). A new measure of agreement of the molecular model to the protein crystal diffraction data was the R_{free} (Brunger 1992) where a 5%–or upto 10% subset of the reflections are excluded from the refinement to secure an unbiased model. Interestingly, chemical crystallography has not adopted the R_{free} statistical indicator. Another measure of reliability is the correlation coefficient. In macromolecular crystallography the quality of the anomalous differences, for example, can be assessed by splitting into two halves a diffraction data set binned into thin resolution shells and calculating the correlation coefficient between the anomalous differences within those two half data sets shell by shell. The various statistical indicators used by macromolecular crystallographers are described in detail by Einspahr and Weiss (2012). In chemical crystallography, besides the R factor, various other parameters are checked such as resolution, redundancy, weighting parameters, goodness of fit, and wR2; the differences between the amplitudes are also checked as a further validation tool. [wR2 is similar to R, but refers to squared F-values, i.e. the intensities. This results in wR2 always having a higher value than R.] One should note that the presence of systematic errors and concerns about the goodness of fit parameter have been expressed in a data review of a large number of chemical crystal structures (Henn 2019).

So, we arrive at the present day in this short history. Today amongst all the great opportunities there are some difficulties, or rather boundaries on us; for a recent discussion see Helliwell (2022) which I now précis.

Firstly, to evaluate a refined model as a data re-user of the PDB, the omit difference electron density map is critical and widely used, and in particular for investigating ligands in the deposited model. There are various options for a structural crystallography researcher in this specific domain. This was the focus of a quite intense CCP4bb debate in late 2020/early 2021 (e.g. involving Dale Tronrud, Rob Nicholls, and myself) about Coot's sharpening/blurring map tool, and in particular the debate raised the question 'is hypothesis after the research is known (HARKing) a problem in macromolecular crystallography (MX)?' The fulcrum of this debate was balancing intuition and false preconception. As Nicholls (2017), building on Nicholls et al (2012), states on behalf of intuition: '*Another feature that can help one to gain intuition is map blurring*' (found under 'Map Sharpening' in the 'Calculate' menu in Coot), which involves adding a very high B factor to the density map to give higher weight to the lower-resolution reflections. This can provide evidence for the presence of structure in the crystal that is currently missing from the model, especially in cases where the ligand is particularly flexible. Indeed, viewing different types of density maps can facilitate the extraction of as much information as possible. But there are quite a range of maps available to the researcher today, and not all are reported. This may be of concern when the map chosen most favourably resembles the researcher's preconceptions; a tendency emphasized by Rupp (2010). Types of difference map include:

(1) A 'before adding the ligand at all' type of map, after refinement of the protein alone.

(2) An omit map calculated after having included and refined the ligand in the model, of which there are various types:
   (i) A 'feature enhanced map' (described in Afonine et al 2012).
   (ii) A 'polder map' (described in Liebschner et al 2017).
   (iii) A CCP4 omit map (Winn et al 2011).
   (iv) A Phenix omit map (Afonine et al 2012).
   (v) A composite omit map (Winn et al 2011, Afonine et al 2012). (Interestingly the two software packages take very different times to make this calculation.)

The above options for the choice of difference map are also interrelated with how we estimate the phases of structure factors, initially experimental but then calculated phases, as our model improves. Within this context looms the worry of model bias. Indeed, the concept of an omit map for a ligand seems a flawed one to me, by which I mean that, once included in the model, can one ever remove the worry of that particular model bias. Surely, the best, least biased, difference map is before any ligand is included. That said, a user of a completed model in the PDB needs an omit map to evaluate the interpretation made by the depositors of that model.

Secondly, at the most basic level, when we deposit our crystal structure molecular model at the PDB, we must offer a single model. The crystal structure, though, is likely to include disordered atoms, in both statically diverse and dynamically changing positions. Moreover, there will be displacement waves (phonons) in the crystal (see e.g. Edwards et al (1990)). Crystal perfections and imperfections, as well as macromolecular crystallization, are described in detail in the book by Chayen et al (2010). Within our one model we can, however, readily determine the atomic displacement parameters (ADPs, also known as the atomic B factors) which are usually calculated as isotropic, but, if a sufficiently high diffraction resolution is achievable, then these ADPs can be determined as anisotropic ellipsoids and even modelled by aspherical scattering factors in very favourable cases. The measurement of diffraction data sets at multiple temperatures is a good approach as it allows a determination of the dynamics as distinct from static disorders.

Thirdly, from the researcher point of view, in a given study there may not just be a single experiment, resulting in the one protein model, but also replicates. These may be intentional, replicating chemistry/biochemistry conditions, or may be computational workflow repeats using different software. So, to flood the PDB with those replicates may not be appropriate. However, they can instead be attached to the publication for transparency to the reader of a study. For an example see the time-series of crystal structures of Jacobs et al (2024) for which the end point was deposited in the PDB and the time-series steps on the whole timescale studied were attached to the paper as supplementary data files.

Key Learning Points

- The evolution of chemical crystallography from the 1940s onwards shows how the statistical residual between F_{obs} and F_{calc} was introduced.
- The practice developed of referring to the R_{factor} as a reliability measure.
- Bayesian methodology is particularly important in applying bond length and angle restraints in macromolecular model refinement due to the large number of parameters to be refined and the more restricted number of diffraction data available than in small molecule crystallography.
- Specific portions of a model at the PDB can be checked via omit map calculations of which there are various types.
- Replicates might be appropriate in a study and the coordinates and diffraction data for each of those may flood the PDB, but they can be attached to the publication as well as the main PDB deposition itself.

References

Afonine, P. V., Grosse-Kunstleve, R. W., Echols, N., Headd, J. J., Moriarty, N. W., Mustyakimov, M., Terwilliger, T. C., Urzhumtsev, A., Zwart, P. H., and Adams, P. D. (2012). *Towards automated crystallographic structure refinement with phenix.refine.* Acta Cryst., D68, 352–367.

Booth, A. D. (1945). *Accuracy of atomic co-ordinates derived from Fourier synthesis.* Nature, Lond., 156, 51–52.

Booth, A. D. (1946). *The accuracy of atomic co-ordinates derived from Fourier series in X-ray structure analysis.* Proc. Roy. Soc. London Ser. A, 188, 77–92.

Booth, A. D. (1947). *The accuracy of atomic co-ordinates derived from Fourier series in X-ray structure analysis. III.* Proc. Roy. Soc. London Ser. A, 190, 482–489.

Brunger, A. (1992). *Free R value: a novel statistical quantity for assessing the accuracy of crystal structures,* Nature 355, 472–475.

Chayen, N. E., Helliwell, J. R., and Snell, E. H. (2010). *Macromolecular Crystallization and Crystal Perfection,* IUCr Monographs on Crystallography, No. 24. Oxford University Press, Oxford.

Cruickshank, D. W. J. (1952). *On the relations between Fourier and least-squares methods of structure determination.* Acta Cryst., 5, 511–518.

Cruickshank, D. W. J. (1960). *The required precision of intensity measurements for single-crystal analysis.* Acta Cryst., 13, 774–777.

Cruickshank, D. W. J. (1999). *Remarks about protein structure precision.* Acta Cryst., D55, 583–601.

Edwards, C., Palmer, S. B., Emsley, P., Helliwell, J. R., Glover, I. D., Harris, G. W., and Moss, D. S. (1990). *Thermal motion in protein crystals estimated using laser-generated ultrasound and Young's modulus measurements.* Acta Cryst., A46, 315–320.

Einspahr, H.M. and Weiss, M. (2012). *Quality indicators in macromolecular crystallography: definitions and applications.* Chapter 2.2 *in International Tables for Crystallography Volume F*, 2nd Edition, edited by Arnold, E., Himmel, D. M., and Rossmann, M. G., Wiley, Chichester.

Giacovazzo, C., Monaco, H. L., Artioli, G., Viterbo, D., Milanesio, M., Gilli, G., Gilli, P., Zanotti, G., Ferraris, G., and Catti, M. (2011). Fundamentals in Crystallography, 3rd Edition. *International Union of Crystallography Texts on Crystallography.* Oxford Science Publications, Oxford.

Gurusaran, M., Shankar, M., Nagarajan, R., Helliwell, J. R., and Sekar, K. (2014). *Do we see what we should see? Describing non-covalent interactions in protein structures including precision.* IUCrJ, 1 (1), 74–81. doi: https://doi.org/10.1107/S2052252513031485

Helliwell, J. R. (2022). *Raw diffraction data are our ground truth from which all subsequent workflows develop.* Acta Cryst., D78, 683–689.

Henn, J. (2019). Metrics for crystallographic diffraction-and fit-data: A review of existing ones and the need for new ones. Crystallography Reviews, 25 (2), 83–156.

Hughes, E. W. (1940). *The crystal structure of dicyandiamide.* J. Am. Chem. Soc., 62, 1258–1267.

Hughes, E. W. (1941). *The crystal structure of melamine.* J. Am. Chem. Soc., 63, 1737–1752.

Jacobs, F. J. F., Helliwell, J. R., and Brink, A. (2024). *Time-series analysis of rhenium(I) organometallic covalent binding to a model protein for drug development.* IUCrJ, 11, 359–373.

Konnert, J.H. (1976). *A restrained-parameter structure-factor least-squares refinement procedure for large asymmetric units.* Acta Cryst., A32, 614–617.

Konnert, J.H. and Hendrickson, W.A. (1980). *A restrained-parameter thermal-factor refinement procedure.* Acta Cryst., A36, 344–350.

Kumar, K. S. D., Gurusaran, M., Satheesh, S. N., Radha, P., Pavithra, S., Thulaa Tharshan, K. P. S., Helliwell, J. R., and Sekar, K. (2015). *Online_DPI: A web server to calculate the diffraction precision index for a protein structure.* J. Appl. Cryst., 48, 939–942.

Liebschner, D., Afonine, P. V., Moriarty, N. W., Poon, B. K., Sobolev, O. V., Terwilliger, T. C., and Adams, P. D. (2017). *Polder maps: Improving OMIT maps by excluding bulk solvent.* Acta Cryst., D73, 148–157.

Murshudov, G. N., Vagin, A. A., and Dodson, E. J. (1997). *Refinement of macromolecular structures by the maximum-likelihood method.* Acta Cryst., D53, 240–255.

Nicholls, R. A. (2017). *Ligand fitting with CCP4.* Acta Cryst., D73, 158–170.

Nicholls, R. A., Long, F., and Murshudov, G. N. (2012). *Low-resolution refinement tools in REFMAC5.* Acta Cryst., D68, 404–417.

Rupp, B. (2010). *Biomolecular Crystallography: Principles, Practice and Application to Structural Biology.* Garland Press, Abingdon.

Wilson, A. J. C. (1950). *Largest likely values for the reliability index.* Acta Cryst., 3, 397–398.

Winn, M. D., Ballard, C. C., Cowtan, K. D., Dodson, E. J., Emsley, P., Evans, P. R., Keegan, R. M., Krissinel, E. B., Leslie, A. G. W., McCoy, A., McNicholas, S. J., Murshudov, G. N., Pannu, N. S., Potterton, E. A., Powell, H. R., Read, R. J., Vagin, A., and Wilson, K. S. (2011). *Overview of the CCP4 suite and current developments.* Acta Cryst., D67, 235–242.

4
Mass spectrometry

4.1 History

W L Bragg in undertaking the first X-ray crystal structure showed that mass played a pivotal role. Quoting from page 30 of his book (Bragg 1975):

> '*The First Complete Analyses: The Alkali Halides*
> *I assigned the structure of Fig 12* (see Figure 4.1) *to the alkaline halides in a paper read to The Royal Society in June 1913* [W L Bragg (1913)] .... *These were the first crystals to be analysed by X-rays. As the structure was now established, it was possible to calculate dimensions from the crystal density and the mass of the NaCl molecule. Half a molecule is associated with each small cube of side a = AB in Fig 12* (Figure 4.1) *so*
>
> $$\frac{1}{2}Mm = \rho a^3$$
>
> *where M is the molecular weight, m is the mass of the hydrogen atom, and ρ is the density of the crystal. This gave a value for a of* 2.8×10^{-8}*cm and so established a scale for the measurement of all X-ray wavelengths and crystal spacings.*'

W L Bragg's father William Henry Bragg commented as follows [W H Bragg (1913)]:

> '*From the work now described by W L Bragg it appears that the reflection phenomena lead to a more definite knowledge of crystal structure, and we may now complete various quantitative determinations. ... (namely) the (X-ray) wavelengths of various homogeneous rays as soon as their angles of reflection are known (from an NaCl or other single crystal).*'

The unit cell parameter for the cubic NaCl was needed, therefore, having been established ingeniously from the mass of a crystal, its volume, and the atomic weights of sodium and chlorine, as described above.

Precision and Accuracy in Biological Crystallography, Diffraction, Scattering, Microscopies, and Spectroscopies.
John R. Helliwell, Oxford University Press. DOI: 10.1093/9780198952848.003.0004

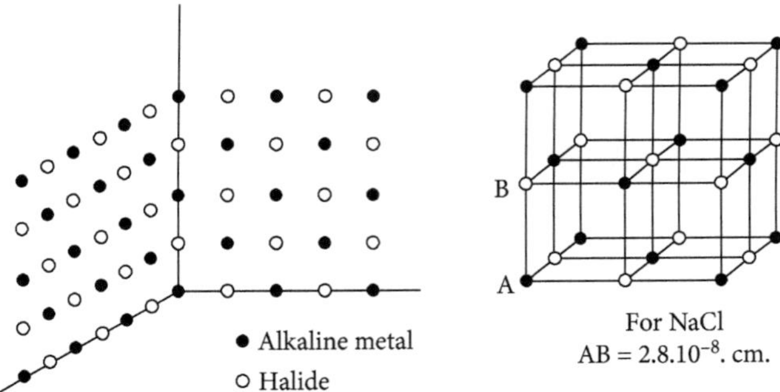

Figure 4.1 The first X-ray crystal structure, NaCl. [The caption shown is the original: See W L Bragg 1913.] The figure published by W L Bragg is reproduced in full showing the layout of both sodium and chlorine atoms in a unit cell. [NB at the time of publication in 1913 it was not yet known that these were sodium ions and chloride ions rather than atoms.]

4.2 Protein crystallography and the Matthews coefficient

The role of mass is very useful in protein crystallography structure determination as exemplified by the paper by Matthews (1968) who showed a simple way to establish the probable number of protein molecules in the crystallographic asymmetric unit. He based his analysis on the available 116 protein crystal structures of the time and found that the most commonly observed values of volume per molecular weight (V_M) are near 2.15 $Å^3$/Dalton, with the median being at 2.61 $Å^3$/Dalton. Thus, for an unknown protein crystal structure, but of known molecular weight and known unit cell dimensions, its likely number of copies in the unit cell can be estimated. Matthews (1968) developed his analysis to lead to the most probable solvent fraction of a unit cell, which was 43%, within a range at that time of 27% to 65%.

4.3 Mass spectrometry modern developments

Mass spectrometry has expanded enormously in its precision, and its applications in analytical chemistry are now routine (Ebsworth et al (1991)). Its role in structural biology has also developed strongly. Liko et al (2016) document a range of critical examples where the stoichiometry of large complexes was established and also chart the critical steps towards its acceptance as a critical method in structural biology.

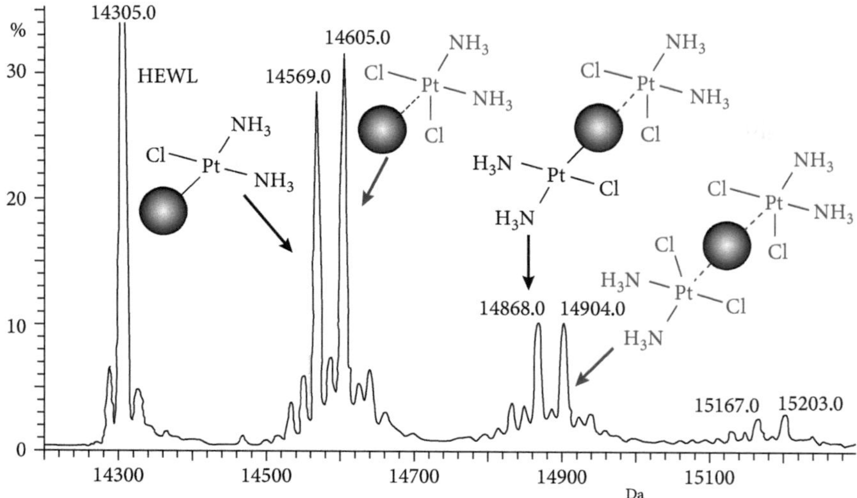

Figure 4.2 Deconvoluted ESI MS spectrum of adducts formed between hen egg white lysozyme (HEWL) and cisplatin, after 48 h incubation at 37 °C. The initial platinum/protein stoichiometry of each sample was 3:1. This figure is reproduced from Casini et al *Biophysical characterisation of adducts formed between anticancer metallodrugs and selected proteins: new insights from X-ray diffraction and mass spectrometry studies.*

From J Inorg Biochem. 2008 May–Jun, 102(5–6), 995–1006 with the permission of Professor Dr Angela Casini and Elsevier.

An example of characterization of anti-cancer covalent bound protein complexes by mass as a complement to X-ray crystallography is provided by Casini et al (2008) and their Figure 4.2. The results depicted in Figure 4.2 would make a precise characterization of what is bound at any one site; if a mixture, a challenging one for X-ray protein crystallography alone.

There are sources of imprecision, however, such as in crystallization of a protein whose mass may be measured prior to crystallization. Note that one can wash and dissolve crystals for mass spectroscopy analysis after the fact, which is a way to verify what was crystallized.

As Casini et al (2008) also remark:

> *Real cellular systems are of course extremely more complicated as they may contain thousands of different proteins, at highly variable concentrations, and also a conspicuous number of low molecular weight components, most of them in relatively high concentration. In addition, these various components are highly organised and compartmentalised within the various intracellular structures thus offering a peculiar spatial distribution inside cells. It is obvious that*

studying the interaction of metallodrugs with proteins inside cells represents today a formidable challenge for researchers.

This quotation of text above is reproduced from Casini et al *Biophysical characterisation of adducts formed between anticancer metallodrugs and selected proteins: new insights from X-ray diffraction and mass spectrometry studies.* J Inorg Biochem. 2008 May–Jun, 102(5–6), 995–1006 with the permission of Elsevier and the author.

The extracts from the papers by W L Bragg and W H Bragg being more than 70 years old do not require permission from the journal, The Royal Society as publisher has informed me.

Key Learning Points

- From the very first crystal structure of the alkali halides by Lawrence Bragg, mass was an important parameter in crystal structure analyses.
- In protein crystal structure analysis, the molecular weight in the unit cell was surveyed for the first known structures in the PDB by Brian Matthews and he made a clear link to the percentage of solvent in the crystal.
- Analytical applications of mass spectrometry are sometimes used as an adjunct to crystallography and can indicate problems for a precise protein crystal structure determination, such as in the case study of the platins described by Angela Casini.
- The research and development of mass spectrometry as an analytical method by Professor Dame Carol Robinson have greatly expanded its scope in structural biology applications.

References

Bragg, W. L. (1913). *The structure of some crystals as indicated by their diffraction of X-rays.* Proc. R. Soc. Lond. A, 89, 248–277.

Bragg, W. L. (1975). *The Development of X-ray Analysis.* Dover Publications, New York.

Bragg, W. H. (1913). *The Reflection of X-rays by Crystals. (II.).* Proc. R. Soc. Lond. A, 89, 246–248.

Casini A., Guerri A., Gabbiani C., and Messori L. (2008). *Biophysical characterisation of adducts formed between anticancer metallodrugs and selected proteins: new insights from*

X-ray diffraction and mass spectrometry studies. J. Inorg. Biochem., May–Jun, 102 (5–6), 995–1006. doi: 10.1016/j.jinorgbio.2007.12.022. Epub 2008 Jan 8. PMID: 18289690.

Ebsworth, E. A. V., Rankin, D. W. H., and Cradock, S. (1991). *Structural Methods in Inorganic Chemistry*, 2nd edition. Blackwell, Oxford.

Liko, I., Allison, T. M., Hopper, J. T. S., and Robinson, C. V. (2016). *Mass spectrometry guided structural biology.* Curr. Opin. Struct. Biol., 40, 136–144.

Matthews, B. W. (1968). *Solvent content of protein crystals.* J. Mol. Biol., 33(2), 491–497.

5
Structure validation approaches

Firstly, let's recall a bit of history. The foundational milestone for protein crystallography was the award of the 1962 Nobel Prize for Chemistry to Max Perutz and John Kendrew (https://www.nobelprize.org/prizes/chemistry/1962/summary/). The role of Dorothy Hodgkin in the determination of crystal structures of biological substances such as penicillin and vitamin B12 and in protein crystallography was recognized by the Nobel Prize for Chemistry in 1964 (https://www.nobelprize.org/prizes/chemistry/1964/summary/). Dorothy Hodgkin's approach of a clear integration of small and large molecule crystal structure studies is especially notable and important to this day.

Thereafter, the history of protein crystallography is basically a one-to-one correspondence with that of the Protein Data Bank (PDB), founded in 1971 with seven protein structures (Bernstein et al (1971)). The PDB provides an organized open access repository to depositor and user of the data held there. A historical perspective of the PDB has been described by Helen Berman (Berman (2008)): '*the establishment of the Protein Data Bank archive, jointly operated by the CCDC and BNL, was announced in Nature New Biology (Protein Data Bank, 1971)*'. [CCDC = Cambridge Crystallographic Data Centre and BNL = Brookhaven National Laboratory.]

In crystallography, starting with coordinates then adding structure factors, we are now making an effort to extend linking of research articles to the primary experimental data (Helliwell et al (2024) provide an overview). The PDB has now introduced versioning (https://www.emdataresource.org/news/keep_pdb_id.html) whereby, quoting from the PDB's website:

> *versioning now allows depositors to update their coordinates while retaining the same PDB accession code. Initially, versioning is limited to PDB entries that were submitted via the OneDep system, which was introduced in 2014 (with deposition for 3DEM methods added in 2016). Eventually functionality will be extended to entries deposited in the legacy systems.*

The PDB also plays a great role in training of researchers in its validation methods and deposition tools including a well-detailed website and offering

Precision and Accuracy in Biological Crystallography, Diffraction, Scattering, Microscopies, and Spectroscopies. John R. Helliwell, Oxford University Press. © John R. Helliwell (2025). DOI: 10.1093/9780198952848.003.0005

training webinars as well as workshops at conferences. It also undertakes a lot of public outreach educational activities with its 'molecule of the month' and its calendars. This weblink http://www.wwpdb.org/validation/validation-reports is a useful entry point and the links within this page present the user guides for the validation reports for each experimental method.

In this book I argue that we need to pay more attention to the accuracy as well as precision. If a PDB deposit is accompanied by a publication, although around 20,000 are not, then one must hope that the structure did indeed have something to say as described in the publication narrative linked with it and that a structure will be corroborated by other methods, and thereby a study will be accurate.

Refereeing of publications, I see as vital. From my career perspective an epiphany moment was becoming Editor in Chief of *Acta Crystallographica* in 1996, through to 2005. In that role, besides those co-editors in my own area of protein crystallography, I worked closely with co-editors in chemical crystallography amongst others. The tradition in *Acta Crystallographica Section C Crystal Structure Communications*, as it was called then, was that the editor and their chosen referees required not only the article, but also the coordinates, the structure factors, and the checkCIF report (Hall and McMahon (2016)) be provided by the authors. It was in the editor's and referees' scope then to reanalyse the crystal structure before accepting them as the versions of record. These were then published together in *Acta Cryst C* as a crystal structure communication. This to me, as a trained physicist, ensured the 'most precise possible' communication of a crystal structure study in its narrative and its data. These versions of record could finally be harvested from *Acta Cryst C* with confidence by a database.

What was happening in biological crystallography in those nine years? Articles were submitted to *Acta Cryst D: Biological Crystallography* and in parallel the PDB would get the data files, initially the coordinates then extended to structure factors later. In 2002 I attended the IUCr Commission on Biological Macromolecules Open Meeting held at the IUCr Congress in Geneva. I spoke as Editor in Chief to an agenda item where I was proposing that the *Acta Cryst C* article with data submission refereeing procedure be adopted for *Acta Cryst D* article submissions. To my dismay that proposal was rejected. Trying to effect a policy change was clearly not going to be easy. In any case, my three terms totalling nine years as Editor in Chief concluded in 2005. One legacy contribution towards improved precision of a protein structure deposition was our IUCr journals' efforts with the PDB to instigate a validation report, akin to the checkCIF report in chemical crystallography. This was christened in 2010 as the PDB Validation Report [https://www.wwpdb.org/validation/2016/XrayValidationReportHelp].

During 2021, IUCr's biological journals adopted the submission procedure whereby coordinates and structure factors are available to referees and editors as well as the PDB Validation Report. This is a great step forward. The impact in terms of the quality and depth of a peer reviewer's report could also benefit from accessing the raw diffraction data, as well as the processed and derived data. Raw data zip files are too large for download to all parts of the globe where a referee might live, so extending the availability of raw data to referees probably has to wait a while.

In my article on data science skills for biological X-ray crystallography referees (Helliwell (2018)) I describe a range of checks that a referee can undertake with the to-be-released PDB data files along with the checks to be made by the referee of the PDB Validation Report. The referee can check what has not made it into the authors' model and the precision of what has been modelled by the authors. The raw diffraction data reprocessing can also be requested if the processed diffraction data described by authors may still have more useful information to be extracted. A common type of request to authors is to more carefully scrutinize their chosen resolution limit within their existing diffraction images. In particular, I think they should adopt the 'paired refinement method' (Karplus and Diederichs (2012)), neatly implemented in PDB REDO (Joosten et al (2014), Maly et al (2020)). Another type of request is the need for new raw data involving a different choice of X-ray wavelength to manipulate the anomalous dispersion signal to properly establish the identity of metal atoms at specific sites in the 3D structure (see Chapter 7). Also, an X-ray fluorescence scan at the beamline establishes quite what metal atoms are actually present in the sample (Leonard et al (2009)).

Post-publication peer review is a trend. These have been both at the level of improving the precision of deposited structures and of correcting inaccuracies in asserted structure–function relationships (Rupp et al (2016)). It seems to me self-evident that the need for these post-publication critique studies would have been greatly reduced if pre-publication peer review of articles included examination of the data. Obviously, it puts more effort and onus onto the pre-publication peer reviewers; an alternative procedure to be adopted by a journal may be adding an extra 'data reviewer', which is an option at the Royal Society of Chemistry journals where chemical crystal structures are involved.

Post-publication reanalyses of the metal sites in metalloprotein structures have led to improvements in existing PDB depositions. Using ion beam analysis through particle-induced X-ray emission (PIXE), Grime et al (2020) quantitatively identified the metal atoms in 30 previously structurally characterized proteins. Over half of these metals had been misidentified in the deposited structural models and, by replacing with the correct metal, the precision of the structural models was improved. The metals reported on were restricted to transition metals. Of course, a pre-publication referee's report can request remeasurement if the identity of a metal is insufficiently proven or the bond lengths and angles do not match expected dictionary values. In another study (Djinović-Carugo and Carugo (2019)), a bioinformatics survey revealed metal cations swimming in the solvent channels of the protein crystals. This required regenerating the protein crystal packings in their post-publication peer review.

Parts of this chapter I published in Helliwell, J. R. (2022). Pre- and post-publication verification for reproducible data mining in macromolecular crystallography. In: Carugo, O., Eisenhaber, F. (eds) *Data Mining Techniques for the Life Sciences. Methods in Molecular Biology, vol 2449*. Humana Press, Springer Nature, New York, NY. https://doi.org/10.1007/978-1-0716-2095-3_10.

Key Learning Points

- Pre-publication validation has emerged as an essential approach for databases and journals to ensure as much as possible that correct versions of record of articles and their underpinning data files (coordinates and structure factors) are available.
- General checks are possible for chemical and macromolecular crystallography and are embedded in web-enabled tools for authors/depositors such as checkCIF and the PDB Validation Report respectively.
- Specialized checks for particular structures mean that consultation by an editor of their chosen specialist referees able to scrutinize these is essential.
- Post-publication peer review sometimes expedites corrections to depositions and their associated articles. Likewise, any such peer review critique must be peer reviewed by a journal, as misinformation can arise as much from an incorrect critique as from an incorrect article.
- Preprints, being a relatively recent phenomenon compared with many journals, are in a state of evolution where they should seek to match the validation and peer review scrutiny of journal articles and associated depositions.

References

Bernstein, F. C., Koetzle, T. F., Williams, G. J., Meyer Jr, E. F., Brice, M. D., Rodgers, J. R., Kennard, O., Shimanouchi, T., and Tasumi, M. (1977). *The Protein Data Bank. A computer-based archival file for macromolecular structures.* Eur. J Biochem., 80, 319–324; and Protein Data Bank (1971), Nature New Biol., 233, 223.

Berman, H. M. (2008). *The Protein Data Bank: A historical perspective.* Acta Cryst A, 64, 88–95.

Djinović-Carugo, K., and Carugo, O. (2019). *Naked metal cations swimming in protein crystals.* Crystals, 9 (11), 581.

Grime, G. W., Zeldin, O. B., Snell, M. E., Lowe, E. D., Hunt, J. F., Montelione, G. T., Tong, L., Snell, E. H., and Garman, E. F. (2020). *High-throughput PIXE as an essential quantitative assay for accurate metalloprotein structural analysis: Development and application.* Journal of the American Chemical Society, 142 (1), 185–197.

Hall, S. R. and McMahon, B. (2016). *The implementation and evolution of STAR/CIF ontologies: Interoperability and preservation of structured data.* Data Science Journal, 15, 3. doi: http://doi.org/10.5334/dsj-2016-003.

Helliwell, J. R. (2018). *Data science skills for referees: I Biological X-ray crystallography.* Crystallography Reviews, 24, 263–272.

Helliwell, J. R., Hester, J. R., Kroon-Batenburg, L. M. J., McMahon, B., and Storm, S. L. S. (2024). *The evolution of raw data archiving and the growth of its importance in crystallography.* IUCrJ, 11, 464–475.

Joosten, R. P., Long, F., Murshudov, G. N., and Perrakis, A. (2014). *The PDB_REDO server for macromolecular structure model optimization.* IUCrJ, 1, 213–220.

Karplus, P. A. and Diederichs, K. (2012). *Linking crystallographic model and data quality.* Science, 336(6084), 1030–1033. doi: 10.1126/science.1218231

Leonard, G. A., Solé, V. A., Beteva, A., Gabadinho, J., Guijarro, M., McCarthy, J., Marrocchelli, D., Nurizzo, D., McSweeney, S., and Mueller-Dieckmann, C. (2009). *Online collection and analysis of X-ray fluorescence spectra on the macromolecular crystallography beamlines of the ESRF.* J. Appl. Cryst., 42, 333–335.

Maly, M., Diederichs, K., Dohnálek, J., and Kolenko, P. (2020). *Paired refinement under the control of PAIREF.* IUCrJ, 7, 681–692.

Rupp, B., Wlodawer, A., Minor, W., Helliwell, J. R., and Jaskolski, M. (2016). *Correcting the record of structural publications requires joint effort of the community and journal editors.* The FEBS Journal, 283, 4452–4457.

6
Other validation tools
Round-robin projects

Measurements on a number of instruments owned by different laboratories, with different intrinsic characteristics and designs, can be made in what is called 'a round-robin study'. These can assess the performance and the limits of the current state of the art of a given method. These types of projects are effective at checking good practice between different laboratories in any analytical method that has become established. The International Union of Crystallography has, through its Commissions, organized several. As representative examples see for instance, EXAFS (Chantler et al (2018)), biological small angle X-ray and neutron scattering, SAXS and SANS (Trewhella et al (2022)), X-ray reflectometry (Colombi et al (2008)), materials science small angle X-ray scattering (Pauw et al (2023)), SANS of polystyrene latex of a standard radius (720 Å) (Rennie et al (2013)), and a very early example of protein crystallography, raw X-ray diffraction data processing with a single data set and multiple UK laboratories taking part (Helliwell et al (1981)). An abstract of an ongoing round-robin project for inorganic and organic electron crystallography is that of Gemmi et al (2023).

In all round-robin projects, a challenging aspect is having standardized samples for measurement. This is especially true for biological samples. An excellent description for biological solution SAXS and SANS is in Trewhella et al (2022). Helliwell et al (1981) side-stepped the need for a standard protein crystal by having an already measured raw diffraction data set. This comprised a collection of oscillation X-ray films measured from a single crystal of apoferritin on a home laboratory X-ray source, which were passed from one UK laboratory to another. Possibly the simplest standard sample is the first of the three different types chosen for investigation in the EXAFS round robin of Chantler et al (2018), which was of a pure metal foil of copper purchased from Goodfellow metals. Figure 6.1 shows a basic example of what could be learnt in this case, where use was made of two different vertical slit settings of an instrument and their comparison with a reference library spectrum (Chantler et al (2018)).

Precision and Accuracy in Biological Crystallography, Diffraction, Scattering, Microscopies, and Spectroscopies.
John R. Helliwell, Oxford University Press. © John R. Helliwell (2025). DOI: 10.1093/9780198952848.003.0006

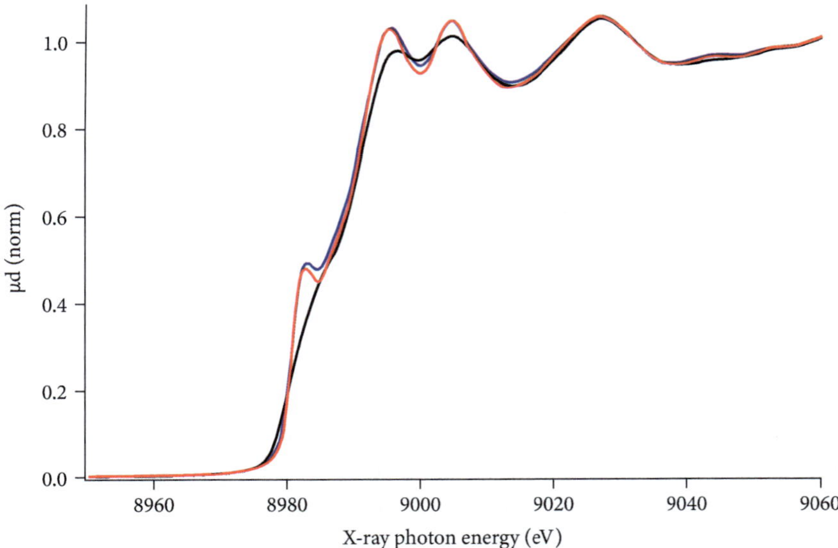

Figure 6.1 Three copper metal foil spectra, two of which were measured at the beamline P65 at PETRA III using different vertical slit sizes (blue, black); the third one (red), which shows most spectral detail, is taken from the EXAFS spectra library (https://cars.uchicago.edu/xaslib/spectrum/606). All three were measured at room temperature using Si (111) double-crystal monochromators. The *x*-axis (horizontal) is labelled in photon energy (electron volt) units, and the *y*-axis is the absorption. The comparison with an externally measured reference spectrum clearly demonstrates that even the blue spectrum is broadened by suboptimal energy resolution.
From Chantler et al (2018) with the permission of Professor Chris Chantler and the IUCr Journals.

Another challenging aspect of round-robin projects is the human factor (Pauw et al (2023)).

Furthermore, Chantler et al (2018) neatly explain the roles envisaged for their planned round robin, stating that:

With the rapid growth of synchrotrons and XAFS-related beamlines, attention has swung to the consistency of data from different beamlines and the cross-portability of data, with respect to data formats, data collection, data content and beamline dependencies. A user may want to know the quality of the data collected on a given day; the beamline scientist will want to rely upon the standard protocols to a known level of accuracy or insight; the management will want to confirm the outstanding results obtained in the laboratories and beamlines under their control. Due to the high monochromatic flux at most EXAFS beamlines around the world and hence high statistical precision per data set, the observed variance between spectra measured at different places must be due to factors other than the purely statistical (Chantler et al (2012)).

The above text quote is from Chantler et al (2018) with the permission of Professor Chris Chantler and the IUCr Journals.

A major lesson from the Helliwell et al (1981) round-robin project was that profile fitting of the diffraction spots gave the most precise estimations, notably of the weaker intensity diffraction spots, rather than the simpler box integration method that was more common at the time due to computer hardware limitations. Profile fitting, of course, was steadily incorporated into raw diffraction data processing software as computer hardware capabilities steadily improved.

A major lesson learnt in the Trewhella et al (2022) study was that inline size exclusion chromatography at the beamline was especially effective in avoiding macromolecular aggregation or interparticle interference.

How do we separate random errors and systematic errors effects in scrutinizing the variations seen in round-robin studies? This is clearly documented in the Rennie et al (2013) round robin of SANS on polystyrene latex spheres study, who commented as follows: -

Scattering from monodisperse, uniform spherical particles is simple to calculate and displays sharp minima. Such data test the calibrations of intensity, wavelength and resolution as well as the detector response. Smoothing due to resolution, multiple scattering and polydispersity has been determined. Sources of uncertainty are often related to systematic deviations and calibrations rather than random counting errors. The study has prompted development of software to treat modest multiple scattering and to better model the instrument resolution. These measurements also allow checks of data reduction algorithms and have identified how they can be improved. The reproducibility and the reliability of instruments and the accuracy of parameters derived from the data are described.

Text extract reproduced from Rennie et al (2013) with the permission of Professor Adrian Rennie and the IUCr Journals.

Furthermore, the results from the SANS can be readily compared with microscopy images (see Figure 6.2).

The National Institute of Standards and Technology (NIST) agency of the U.S. Department of Commerce is responsible for the standardization of

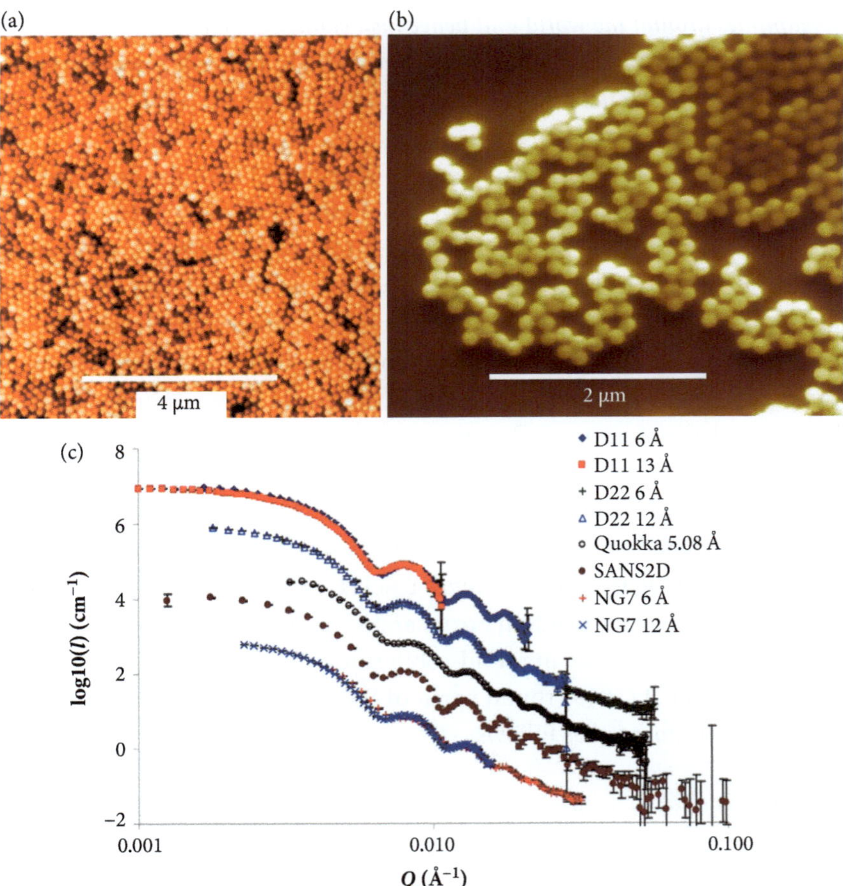

Figure 6.2 Comparison of data from three sources. (*a*) Atomic force microscopy and (*b*) scanning electron microscopy images of PS3 polystyrene latex. Both images show that the particles are spherical and have mean diameters of about 1450 Å, but the resolution does not allow a precise determination of the size distribution. (*c*) Comparative SANS data for the PS3 sample A from various instruments (for each instrument the data are offset by a factor of 10 for clarity). As an example: For ILL D22 SANS the particle radius was fitted as 724 Å with a Gaussian size distribution with standard deviation of 20 Å. Data for SANS2D were not scaled. The *y*-axis is of the logarithm of the scattered intensities and the *x*-axis is the scattering Q value in the units of $Å^{-1}$.
From Rennie et al (2013) with permission of Professor Adrian Rennie and the IUCr Journals.

weights and measures, timekeeping, and navigation. Thus, it is where one can purchase a standard sample of silicon powder which is useful at a diffraction beamline to calibrate its X-ray wavelength. This calibration can also be done for a synchrotron beamline by a tuneable photon energy scan across a pure metal foil. A home laboratory X-ray source is based on an element emission wavelength such as Cu K α (=1.54Å), actually a doublet of two wavelengths $K_{\alpha 1}$ (=1.54056 Å) and $K_{\alpha 2}$ (=1.54439 Å). The X-ray wavelength is needed to precisely determine the unit cell of a crystal.

A challenge not yet overcome for macromolecular X-ray crystallography is a community agreed standard, single, protein crystal as a reference standard for calibration. The desirable characteristics of such a crystal would certainly be stability under X-ray irradiation at a range of incident intensities or at least for a sufficient dose that reasonable data completeness would be possible. It would have to be routinely made available and low cost.

Key Learning Points

- Round-robin projects are a well-established approach to establish good practice in measurements and analysis.
- Various of the IUCr Commissions have organized or endorsed round-robin investigation by experts in their domains.
- Round-robin studies provide a benchmark when working with standard samples for all practitioners to follow according to the circumstances of their particular instrument.

References

Chantler, C. T., Barnea, Z., Tran, C. Q., Rae, N. A., and de Jonge, M. D. (2012). *A step toward standardization: Development of accurate measurements of X-ray absorption and fluorescence.* J. Synchrotron Rad., 19, 851–862.

Chantler, C. T., Bunker, B. A., Abe, H., Kimura, M., Newville, M., and Welter, E. (2018). *A call for a round robin study of XAFS stability and platform dependence at synchrotron beamlines on well-defined samples.* J. Synchrotron Rad., 25, 935–943.

Colombi, P., Agnihotri, D. K., Asadchikov, V. E., Bontempi, E., Bowen, D. K., Chang, C. H., Depero, L. E., Farnworth, M., Fujimoto, T., Gibaud, A., Jergel, M., Krumrey, M., Lafford, T. A., Lamperti, A., Ma, T., Matyi, R. J., Meduna, M., Milita, S., Sakurai, K., Shabel'nikov, L., Ulyanenkov, A., Van der Lee, A., and Wiemer, C. (2008). *Reproducibility in X-ray reflectometry: Results from the first world-wide round-robin experiment.* J. Appl. Cryst., 41, 143–152.

Gemmi, M., Palatinus, L., Klar, P., Faye, M., Agbemeh, V., Hajizadeh, A., Chintakindi, H., Suresh, A., Jeriga, B., Gemmrich Hernández, L., Santucci, M., Wang, L., Vypritskaia, A., Passuti, S., Cordero Oyonarte, E., Matinyan, S., and Ben Meriem, A. (2023). *Round robin results of 3D electron diffraction data on inorganic and organic compounds.* Acta Cryst., A79, C1085.

Helliwell, J. R., Achari, A., Bloomer, A. C., Bourne, P. E., Carr, P., Clegg, G. A., Cooper, R., Elder, M., Greenhough, T. J., Shaanan, B., Smith, J. M. A., Stuart, D. I., Stura, E. A., Todd,

R., Wilson, K. S., Wonacott, A. J., and Machin, P. A. (1981). *Protein crystal oscillation film data processing: a comparative study.* Acta Cryst., A37, C311–C312. The overheads describing this project presented at the IUCr Ottawa World Congress of Crystallography in 1981 are available from Helliwell, J.R. here: https://zenodo.org/records/166325.

Pauw, B. R., Smales, G. J., Anker, A. S., Annadurai, V., Balazs, D. M., Bienert, R., Bouwman, W. G., Breßler, I., Breternitz, J., Brok, E. S., Bryant, G., Clulow, A. J., Crater, E. R., De Geuser, F., Del Giudice, A., Deumer, J., Disch, S., Dutt, S., Frank, K., Fratini, E., Garcia, P. R. A. F., Gilbert, E. P., Hahn, M. B., Hallett, J., Hohenschutz, M., Hollamby, M., Huband, S., Ilavsky, J., Jochum, J. K., Juelsholt, M., Mansel, B. W., Penttilä, P., Pittkowski, R. K., Portale, G., Pozzo, L. D., Rochels, L., Rosalie, J. M., Saloga, P. E. J., Seibt, S., Smith, A. J., Smith, G. N., Spiering, G. A., Stawski, T. M., Tache, O., Thünemann, A. F., Toth, K., Whitten, A. E., and Wuttke, J. (2023). *The human factor: results of a small-angle scattering data analysis round robin.* J. Appl. Cryst., 56, 1618–1629.

Rennie, A. R., Hellsing, M. S., Wood, K., Gilbert, E. P., Porcar, L., Schweins, R., Dewhurst, C. D., Lindner, P., Heenan, R. K., Rogers, S. E., Butler, P. D., Krzywon, J. R., Ghosh, R. E., Jackson, A. J., and Malfois, M. (2013). *Learning about SANS instruments and data reduction from round robin measurements on samples of polystyrene latex.* J. Appl. Cryst., 46, 1289–1297.

Trewhella, J., Vachette, P., Bierma, J., Blanchet, C., Brookes, E., Chakravarthy, S., Chatzimagas, L., Cleveland, T. E., Cowieson, N., Crossett, B., Duff, A. P., Franke, D., Gabel, F., Gillilan, R. E., Graewert, M., Grishaev, A., Guss, J. M., Hammel, M., Hopkins, J., Huang, Q., Hub, J. S., Hura, G. L., Irving, T. C., Jeffries, C. M., Jeong, C., Kirby, N., Krueger, S., Martel, A., Matsui, T., Li, N., Perez, J., Porcar, L., Prange, T., Rajkovic, I., Rocco, M., Rosenberg, D. J., Ryan, T. M., Seifert, S., Sekiguchi, H., Svergun, D., Teixeira, S., Thureau, A., Weiss, T. M., Whitten, A. E., Wood, K., and Zuo, X. (2022). *A round-robin approach provides a detailed assessment of biomolecular small-angle scattering data reproducibility and yields consensus curves for benchmarking.* Acta Cryst., D78, 1315–1336.

7
Similarities and differences in the probes used in structure determination

The four probes used are X-rays, neutrons, electrons, and NMR.

The atomic scattering factor of atoms for X-rays steadily increases for increasing atomic number. The resonance effects ('anomalous dispersion') provided via the X-ray wavelength tuning to and around an individual element K or L absorption edge give X-ray diffraction an exquisite sensitivity to identifying a metal atom, especially bimetallic cases of neighbouring atomic number. The example with the most X-ray wavelengths, 11, to investigate a zinc-substituted gallium phosphate involving partial occupancy, is that of M. Helliwell et al (2010). By tuning to each of the zinc and gallium K absorption edges a very precise crystal structure was determined.

This is a long-standing challenge for which I have argued tuneable synchrotron radiation is ideal (Helliwell (1984)). Einspahr et al (1985) described the case of discriminating between a manganese ion and a calcium ion in pea lectin. More recent examples are described in Brink and Helliwell (2017) and Pontillo et al (2017), briefly described below. There is significant utility in using two synchrotron X-ray wavelengths for element discrimination or enhanced sensitivity to lower metal atom occupancy; the variation in f'' is shown for simplicity in Figure 7.1. The arrows indicate the X-ray wavelengths used in measuring the diffraction data sets in the studies (Brink and Helliwell (2017) and Pontillo et al (2017)). There is also a variation in f', which must be allowed for in refining a metal atom occupancy. Discriminating between two elements in the cases shown in Figure 7.1a is of Pt versus Cd harnessed in the crystal structure analysis of encapsulating the anti-cancer drug carboplatin in a (ferritin) protein shell nanoparticle; ferritin crystals require cadmium ions for their crystallization but the range of metal occupancies leads to either Pt or Cd being plausible interpretations at binding sites unless two X-ray wavelengths are used (Pontillo et al (2017)). Figure 7.1.(b) shows how to enhance the sensitivity of crystal structure analysis to determine the identity of weak metal binders, where using two X-ray wavelengths

Precision and Accuracy in Biological Crystallography, Diffraction, Scattering, Microscopies, and Spectroscopies.
John R. Helliwell, Oxford University Press. © John R. Helliwell (2025). DOI: 10.1093/9780198952848.003.0007

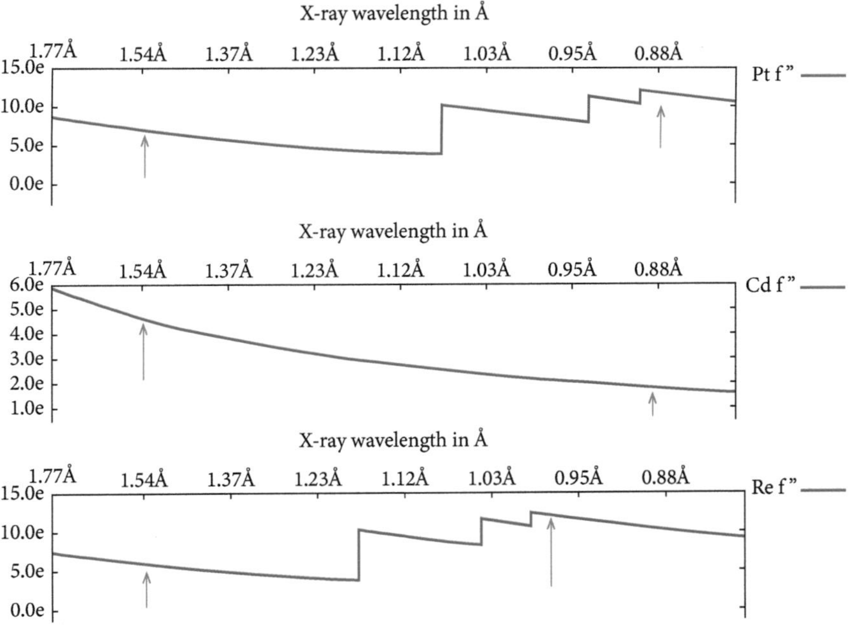

Figure 7.1 Plots of the resonant scattering coefficient f″ in units of electrons along the *y*-axis (vertical) versus X-ray wavelength in Å along the *x*-axis (horizontal). Variation in f″ for the elements of platinum (top), cadmium (middle) and rhenium (bottom). The arrows indicate the X-ray wavelengths used in the studies respectively of Pontillo et al (2017) and of Brink and Helliwell (2017). These figures were prepared using http://skuld.bmsc.washington.edu/scatter/AS_form.html.
Copyright Ethan A. Merritt ©1996–2011 to whom I am very grateful.

allows for a comparison of peak height changes between the two as demonstrated in the case shown for rhenium; see also Table 7.1 (Brink and Helliwell (2017)).

An overview of the principal features of metals in protein structures is given by Harding et al (2010). A database overview of metal coordination distances in metalloproteins is described by Bazayeva et al (2024). Lin et al (2024) developed a database, MESPEUS, derived from an earlier database of the same name made by Harding et al (2010) and references therein. Using PIXE (particle-induced X-ray emission), Snell et al (2023) and references therein have identified 'mythical' metal assignments in the PDB. Clearly, improved procedures for routine assessment of metal content are needed and can be readily achieved at central facility synchrotron radiation macromolecu-

Table 7.1 The increase in anomalous difference electron density peak heights for the example described in Brink and Helliwell (2017) in their Cu Kα and Diamond Light Source (DLS) diffraction data due to the optimized diffraction wavelength. Three examples are shown to illustrate high, medium, and lower occupancy rhenium binding sites.

Residue	Cu Kα ($\lambda = 1.5418$ Å)	DLS ($\lambda = 0.9763$ Å)
	Re $f'' = 5.9$	Re $f'' = 12.1$
His 15	19.3	43.1
Asp 52	6.9	12.9
Glu 35	4.8	12.2

lar crystallography beamlines via X-ray fluorescence. This has been introduced at ESRF (Leonard et al (2009)). In a different approach, Zheng et al (2014) have developed a webserver called 'check my metal'.

The nuclear scattering of different isotopes using neutrons also provides exquisite sensitivity with contrast variation involving isotope pairs such as hydrogen and deuterium. Likewise, there are element-to-element variations, but much less marked than using X-rays. A particularly attractive feature of the neutron-scattering factors approach is that deuterium is as good a scatterer for neutrons as carbon. This is exploited in neutron macromolecular crystallography to determine the protonation states of ionizable amino acid residues (aspartic acid, glutamic acid, histidine, and lysine).

Electrons also show a steadily increasing variation of the electrostatic potential with atomic number. They are much more sensitive to hydrogen atoms than X-rays. For those biological molecules known from chemical analysis to contain an electron-dense metal centre, the location of those metal atoms can be found. For bimetallic cases of closely similar atomic number, the sensitivity is, to my knowledge, untested with electrons. For the ionizable amino acid side chains mentioned above, electron beam damage is a risk, and thus absence of a hydrogen atom may be due to experiment rather than the pK_a of the side chain.

Both electrons and X-rays can potentially change the oxidation state of a metal during the experiment. With X-rays there is the important exception, however, of a femtosecond time range pulsed X-ray laser where the 'diffract before destroy' approach (Neutze et al 2000), makes it possible to directly identify the oxidation state.

7.1 Sample scattering power

Perhaps surprisingly this term 'scattering power' was not introduced as an equation until Andrews et al (1988); see their Table 3 comparing microcrystal diffraction of both chemical and protein microcrystals. A different approach was adopted by Henderson (1995), namely that for naturally occurring biological material, electrons provide the most information for a given amount of radiation damage; the study involved comparing X-rays, neutrons, and electrons. This result concurs with the Vainshtein (1964) approach of comparing the three probes where electron scattering is the strongest, and so the very smallest crystals can be used. The obvious utility of the Henderson (1995) analysis for single particles of biological macromolecules encouraged the development of cryoEM. Once a crystal is grown to above a size of ~0.1 to 1 micron, then the situation changes. Namely, electrons are overtaken by X-rays in their utility for structure analysis. Furthermore, if the crystal grows to ~1mm^3 or more, then neutrons become viable and are by far the best because they yield a complete structure with hydrogens at room temperature and free from radiation damage. Thus, neutron diffraction has great potential for data collection at body temperature (37 °C) (see Jacobs et al (2024)). Table 7.2 compares the fundamentals of X-rays and electrons, listing their pros and cons. Since neutrons are the ideal probe for crystallography, their use is purely one of practical challenges, and these are compared with X-rays and electrons in Table 7.2. There is one fundamental feature of neutron crystallography which is that to analyse a neutron protein crystal structure to complete the protonation details requires a prior X-ray crystal structure.

Tables 7.2 and 7.3, taken from my article, Helliwell (2021), are reproduced here with the permission of IUCr Journals.

7.2 The important role of nuclear magnetic resonance (NMR)

NMR is a method of determining structures of (small) proteins in solution, without crystallization, and for an intrinsically disordered protein is often the method of choice. [NB NMR studies of proteins in solution are described in Chapter 13. Also, cryoEM is solution based, see Chapter 11.] The role of NMR has three additional advantages when combining it with a crystal structure. Firstly, an NMR 'structure' is an ensemble of structural models that meet geometric restraints determined by the NMR experiment. Where the

Table 7.2 Fundamental comparison of X-rays and electrons.

X-rays	Electrons
• Yield electron density • Diffraction efficiency relatively low with respect to electrons • Mature method, well understood and validated (but still needs article with data files and checkCIF report for submission of chemical crystallography articles to IUCr Journals[1]) • Phase problem • Radiation damage effects	• Yield an electrostatic potential map • Diffraction efficiency high, good for very small samples, otherwise strongly absorbed and thereby: • Prone to multiple scattering which can seriously affect bond distances and angles • CryoEM and electron crystallography are still improving their capabilities as methods • No phase problem [2] • Radiation damage effects

[1]That procedure was extended to IUCr's biological journals in 2021.
[2]In protein electron crystallography the phase problem is nicely discussed in Gemmi et al (2019) and, as they summarize, molecular replacement is so far the method applied.

protein polypeptide chain is particularly flexible, and X-ray crystal structure electron density maps are unclear, an NMR ensemble presents a range of possible conformations for the polypeptide. Secondly, the NMR data illuminate the dynamics of the relatively static protein core that appears well-ordered in X-ray crystallographic models. This was highlighted by Wüthrich and Wagner (1975) with respect to aromatic side chains such as phenylalanine, where distinctly different NMR spectra could be seen showing three possibilities; spinning, 180 degrees flip, or static. Thirdly the application of NMR directly on a crystal allows for both the ordered and disordered regions in the crystal to be studied simultaneously. The book *NMR Crystallography* (Harris et al (2009)) surveys a large number of examples of this; Chapter 27 by Middleton (2009) is devoted to structural biology applications. Figure 27.2 of Middleton shows the effects of sample preparation in solid-state NMR, comparing polycrystalline and nanocrystalline, which are 'practically identical', and thirdly, lyophilized ubiquitin protein, which shows poor resolution indicating structural heterogeneity. Middleton (2009) also points out that where a crystal is available but does not diffract, or diffraction is limited, then NMR crystallography could be applied to study protein ligand binding on soaking with ligand. Lewandowski et al (2015) used solid-state NMR to investigate protein and solvent motions of nano- and microcrystalline, fully hydrated, protein 'GB1' at temperatures from 105 to 280 K. GB1 is 'a small globular protein specifically binding to antibodies'. Their study described a hierarchical change in dynamic

Table 7.3 Practical challenges for X-rays, electrons, and neutrons.

X-rays	Electrons	Neutrons
• Radiation damage such as splitting of disulfides, truncation of amino acid side chains, changes to oxidation states of metal atoms/ions • This leads to use of cryo-temperature to (partially) mitigate these effects	• Very strong interaction with matter, an advantage for very thin/small samples such as a single molecule or a nanocrystal • Microcrystals have strong electron beam absorption/multiple scattering and any bigger samples cannot be used with an electron beam, in transmission at least • Careful scrutiny of the error estimates on bond lengths and angles is needed	• Weak flux, so: – Use as broad a bandpass of the emitted neutrons as possible, i.e. Laue diffraction – Use as long a mean wavelength as possible to increase scattering efficiency – Reduce any background scattering so as to maximize signal to noise (i.e. in biology change hydrogens for deuteriums with their ~40× less incoherent scattering) – Grow as big a crystal as possible eg ~1mm^3 (typical range 0.1 to 8mm^3) – Maximize the full exploitation of perdeuteration

behaviour (see especially their Figure 4) showing '*a unified description of the essential conformational energy surface, relating the amplitude and activation of solvent, sidechain, and backbone motions in a hierarchical distribution, as well as unambiguous identification of NMR line broadening at cryogenic temperatures*'. Stepping beyond protein examples, Bryce (2017) provides a topical review of the use of NMR crystallography across a wide range of materials, spanning inorganics and organics as well as a protein. The IUCr established a Commission on NMR Crystallography and Related Methods in 2014 in recognition of all these important possibilities for materials characterization (https://www.iucr.org/iucr/commissions/nmr-crystallography).

To connect with the title of this book, the question arises: is a combination of probes better than the precision of each one as a single study, i.e. does cross-validation clearly lead to accuracy? To answer this, we shall consider each pairing of methodologies in turn. Firstly, this means we must again acknowledge that we are always operating from a position of not knowing the dynamic conformation of the protein in actual living cells. In trying to understand the structural chemistry of biological macromolecules in the environment of the living cell, we accept that anything we do involves killing the living cell, whether *in vitro* or as a microtome slice (thin section). In medical diagnosis there are non-invasive, non-damaging probes such as ultrasound scans of a patient (see e.g. https://www.nibib.nih.gov/sites/default/files/2022-04/Fact-Sheet-Ultrasound.pdf) or MRI (Magnetic Resonance Imaging), albeit with contrast imaging agents administered to the patient. However, similarly non-invasive, non-damaging imaging at an atomic scale within living cells seems inconceivable; perhaps the closest we have yet come is cryoEM tomographic reconstructions of macromolecules on vitrified cells and in cell NMR (Luchinat and Banci (2017)).

Let us return to combining probes to study a crystal structure. A major systematic error associated with X-rays is that of not accurately placing the hydrogens. This is most comprehensively, if still imperfectly, achievable using neutron crystallography on the same crystal. The meaning of 'imperfectly' is that excessive motion of any hydrogen smears out the diffraction signal, the nuclear density. Hydrogens can be replaced by deuterium, and database surveys indicate that this procedure does not affect the structure within the usual diffraction resolution of macromolecular crystallography (Fisher and Helliwell (2008)). Overall, complementary X-ray and neutron crystallography analyses of a single macromolecular crystal provides a more complete picture of the macromolecular structure in a crystal. Neutron diffraction data compensate for the systematic error of the X-rays by locating the hydrogen nuclei experimentally. The outcome is thus more precise and provides a more accurate portrayal of the protein within the crystal. However, we accept that it still may not be an accurate model of the macromolecule in its cellular environment. We also note that the molecular dynamics of a macromolecule are constrained in a crystal.

A similar situation exists when an X-ray crystal structure is paired with a solid-state NMR study of that same crystal. The X-rays give the ordered portion of the crystal structure (minus the hydrogens) and the NMR the disordered portion (Bryce (2017)). A combined X-ray with neutron diffraction with NMR spectroscopy would provide a much more complete picture of the crystal contents. To my knowledge, this combined approach has not been applied to a single crystal of biological macromolecule(s) but developments towards that are occurring (Kovalevskiy et al (2018)). So, from the point of view of the model refinement software Refmac, refinement should be possible if all three different types of experimental data were available. This has been applied in the case of the solution state via SAXS, SANS, and NMR (Delhommel et al (2023)).

Parts of this chapter have been published by Helliwell (2021) and have been reproduced with the permission of IUCr Journals.

Key Learning Points

- The probes of the structure of matter, namely X-rays, electrons, and neutrons for scattering and diffraction as well as NMR, have complementary characteristics for structure determination.
- Avoiding or mitigating radiation damage is a challenge for X-rays and electrons.
- Neutrons and NMR are non-destructive probes but have different restrictions, i.e. maximum unit cell size for neutron crystallography and molecular weight maximum for NMR.
- The presence of metals in a sample can be established by use of X-ray fluorescence or particle-induced X-ray emission (PIXE). The actual positions of metal atoms are determined from the anomalous difference Fourier electron density map.
- Surveys of metalloprotein structures are very informative on the structure and coordination preferences of particular metals. Web-enabled validation tools are available.

References

Andrews, S. J., Papiz, M. Z., McMeeking, R., Blake, A. J., Lowe, B. M., Franklin, K. R., Helliwell, J. R., and Harding, M. M. (1988). *Piperazine silicate (EU 19): the structure of a very small crystal determined with synchrotron radiation.* Acta Cryst., B44, 73–77.

Bazayeva, M., Andreini, C., and Rosato, A. (2024). *A database overview of metal-coordination distances in metalloproteins.* Acta Cryst., D80, 362–376.

Brink, A. and Helliwell, J. R. (2017). *New leads for fragment-based design of rhenium/technetium radiopharmaceutical agents.* IUCrJ, 4, 283–290.

Bryce, D. L. (2017). *NMR crystallography: Structure and properties of materials from solid-state nuclear magnetic resonance observables.* IUCrJ, 4, 350–359.

Delhommel, F., Martínez-Lumbreras, S., and Sattler, M. (2023). *Combining NMR, SAXS and SANS to characterize the structure and dynamics of protein complexes.* Methods in Enzymology, 678, 263–297.

Einspahr, H., Suguna, K., Suddath, F. L., Ellis, G., Helliwell, J. R., and Papiz, M. Z. (1985). *The location of manganese and calcium ion cofactors in pea lectin crystals by use of anomalous dispersion and tuneable synchrotron X-radiation.* Acta Cryst, B41(5), 336–341.

Fisher, S. J. and Helliwell, J. R. (2008). *An investigation into structural changes due to deuteration.* Acta Cryst., A64, 359–367.

Gemmi, M., Mugnaioli, E., Gorelik, T. E., et al (2019). *3D electron diffraction: The nanocrystallography revolution.* ACS Central Science, 5(8), 1315–1329.

Harding, M. M., Nowicki, M. W., and Walkinshaw, M. D. (2010). *Metals in protein structures: A review of their principal features.* Crystallogr. Rev., 16, 247–302.

Harris, R. K., Wasylishen, R. E., and Duer, M. J. (Eds) (2009). *NMR Crystallography.* John Wiley, Chichester, UK.

Helliwell, J. R. (1984). *Synchrotron X-radiation protein crystallography: instrumentation, methods and applications.* Reports on Progress in Physics, 47(11), 1403–1497.

Helliwell, J. R. (2021). *Combining X-rays, neutrons and electrons, and NMR, for precision and accuracy in structure function studies.* Acta Cryst., A77, 173–185.

Helliwell, M., Helliwell, J. R., Kaucic, V., Zabukovec Logar, N., Teat, S. J., Warren, J. E., and Dodson, E. J. (2010). *Determination of zinc incorporation in the Zn-substituted gallophosphate ZnULM-5 by multiple wavelength anomalous dispersion techniques.* Acta Cryst., B66, 345–357.

Henderson, R. (1995). *The potential and limitations of neutrons, electrons and X-rays for atomic resolution microscopy of unstained biological molecules.* Quarterly Reviews of Biophysics, 28, 171–193.

Jacobs, F. J. F., Helliwell, J. R., and Brink, A. (2024). *Body temperature protein X-ray crystallography at 37°C: A rhenium protein complex seeking a physiological condition structure.* Chem. Commun. https://doi.org/10.1039/D4CC04245J.

Kovalevskiy, O., Nicholls, R. A., Long, F., Carlon, A., and Murshudov, G. N. (2018). *Overview of refinement procedures within REFMAC5: Utilizing data from different sources.* Acta Cryst., D74, 215–227.

Leonard, G. A., Solé, V. A., Beteva, A., Gabadinho, J., Guijarro, M., McCarthy, J., Marrocchelli, D., Nurizzo, D., McSweeney, S., and Mueller-Dieckmann, C. (2009). *Online collection and analysis of X-ray fluorescence spectra on the macromolecular crystallography beamlines of the ESRF.* J. Appl. Cryst., 42, 333–335.

Lewandowski, J. R., Halse, M. E., Blackledge, M. and Emsley, L. (2015). *Direct observation of hierarchical protein dynamics.* Science, 348(6234), 578–581.

Lin, G. Y., Su, Y. C., Huang, Y. L., and Hsin, K. Y. (2024). *MESPEUS: A database of metal coordination groups in proteins.* Nucleic Acids Research, 52(D1), D483–D493.

Luchinat, E. and Banci, L. (2017). *In-cell NMR: A topical review.* IUCrJ, 4, 108–118.

Middleton, D. (2009) *Structural Biology.* Chapter 27 in Harris R. K., Wasylishen, R. E., and Duer, M. J. (Eds), *NMR Crystallography.* John Wiley, Chichester, UK.

Neutze, R., Wouts, R., van der Spoel, D., Weckert, E., and Hajdu, J. (2000). *Potential for biomolecular imaging with femtosecond X-ray pulses.* Nature, 406, 752–757.

Pontillo, N., Ferraro, G., Helliwell, J. R., Amoresano, A., and Merlino A. (2017). *X-ray structure of the carboplatin-loaded apo-ferritin nanocage.* ACS Med. Chem. Lett., 8 (4), 433–437.

Snell, E. H., Garman, E. F., Grime, G. W., Cohen, A. E., and Bowman, S. E. J. (2023). *The mythical metal insights on the accuracy of metal identification in structural biology.* Acta Cryst., A79, a281.

Vainshtein, B. (1964). *Structure Analysis by Electron Diffraction.* Pergamon Press, Oxford.

Wüthrich, K. and Wagner, G. (1975). *NMR investigations of the dynamics of the aromatic amino acid residues in the basic pancreatic trypsin inhibitor.* FEBS Lett., 50, 265–268.

Zheng, H., Chordia, M., Cooper, D., et al (2014). *Validation of metal-binding sites in macromolecular structures with the CheckMyMetal web server.* Nat Protoc, 9, 156–170. https://doi.org/10.1038/nprot.2013.172

8
Fibre diffraction

Arguably, the most famous discovery of the 20th century was the DNA double helix; see Figure 8.1 from Watson and Crick (1953). Their paper had six references, and they remarked on the experimental data that might support the structure of Figure 8.1 as follows:

> *The previously published X-ray data [5, 6] on deoxyribose nucleic acid are insufficient for a rigorous test of our structure.*

where references 5 and 6 were Astbury, W. T., *Symp. Soc. Exp. Boil.* 1, Nucleic Acid, 66 (Camb. Univ. Press, 1947) and Wilkins, M. H. F., and Randall, J. T., Biochim. et Biophys. Acta, **10**, 192 (1953). They also made the additional remark in their Acknowledgements:

> *We have also been stimulated by a knowledge of the general nature of the unpublished experimental results and ideas of Dr. M. H. F. Wilkins, Dr. R. E. Franklin and their co-workers at King's College, London.*

They also used prior knowledge as a restraint on their model building which they acknowledge as follows:

> *We are much indebted to Dr. Jerry Donohue for constant advice and criticism, especially on interatomic distances.*

They also state another assumption: restraints i.e. (Bayesian) prior knowledge:

> *If it is assumed that the bases only occur in the structure in the most plausible tautomeric forms (that is, with the keto rather than the enol configurations) it is found that only specific pairs of bases can bond together. These pairs are: adenine (purine) with thymine (pyrimidine), and guanine (purine) with cytosine (pyrimidine).*

These quotations are from Watson and Crick (1953) and are reproduced with the permission of Springer Nature.

Precision and Accuracy in Biological Crystallography, Diffraction, Scattering, Microscopies, and Spectroscopies.
John R. Helliwell, Oxford University Press. DOI: 10.1093/9780198952848.003.0008

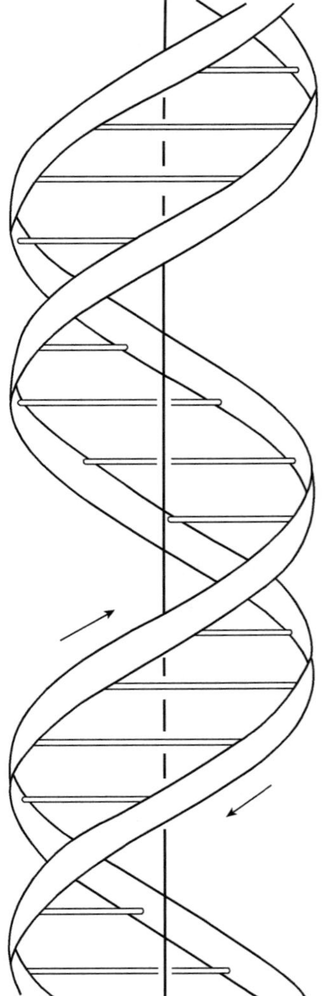

Figure 8.1 A sketch of the 3D double helix structure of deoxyribose nucleic acid (DNA) proposed by Watson and Crick. Horizontal rods represent the pairs of bases holding the chains (vertical) together.

From Watson and Crick (1953) and reproduced with the permission of Springer Nature including the original caption.

Note: the word restraint is employed above but since the modelling was done with mechanical components rather than by computational minimization, the assumptions are actually constraints.

The Watson and Crick (1953) paper and the manner of its communication has been the subject of various books, not least stimulated by the narrative book by J D Watson entitled *The Double Helix*. The most recent of these books

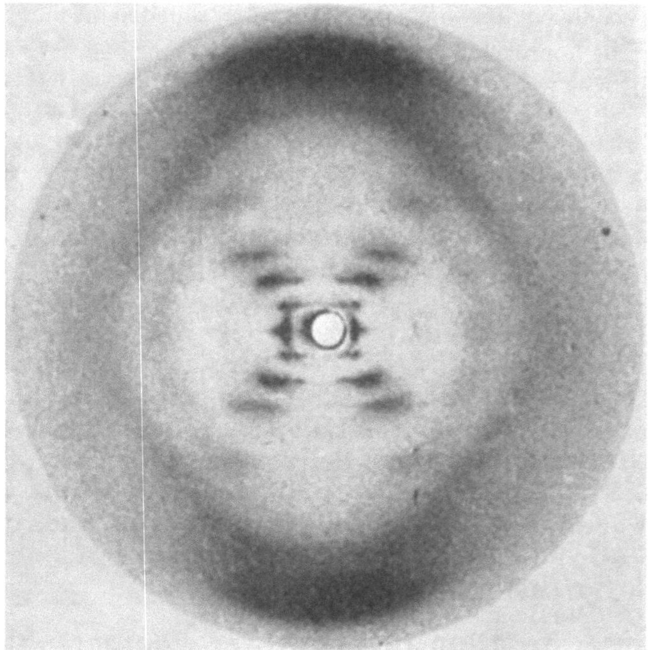

Figure 8.2 A reproduction of the X-ray diffraction photograph taken from a DNA fibre 'structure B' by Franklin and Gosling (1953). It is in greyscale format, black being the most intense parts of the diffraction and light grey the weakest. Reproduced with the permission of Springer Nature including the original caption.

was reviewed by Emeritus Professor Watson Fuller, a PhD student with M H F Wilkins (Fuller (2019)).

A paper not in the references list of Watson and Crick (1953) is the earlier paper coauthored by Crick namely Cochran, Crick, and Vand (1952) which provided the theoretical basis for knowing the Fourier transform of a helix of atoms. There are numerous YouTube teaching demonstrations illustrating what the diffraction of a helix is by using a laser light beam incident onto a helix. These videos illustrate the general likeness to the X-ray diffraction photograph from a DNA fibre (Franklin and Gosling (1953), see Figure 8.2). A comprehensive mathematical treatment of diffraction covering these aspects is by Hukins (1981).

From a modern vantage point and with respect to the title of this book the B form DNA fibre diffraction pattern is less detailed than the A form, which arose from a drier DNA fibre. This latter, it has been extensively discussed, became more the focus of Rosalind Franklin due to its likely better precision when matched to the molecular structure of the DNA. The B form, however, with its much less detailed diffraction pattern, was better suited to the purely mechanical modelling approach adopted by Watson and Crick (1953), who clearly stated that their model was not rigorously supported by the experimental diffraction data.

Study of the transitions between the different forms of DNA using X-ray diffraction have continued into the modern era. See for example Mahendrasingam et al (1986).

An excellent survey of textile fibres and their X-ray diffraction patterns are available in the book of that title by Astbury (1943). This book is probably not so easy to locate but a review of it by Rawlins (1943) is available.

Key Learning Points

- The double helix structure of DNA is a good case study to learn from.
- Fibre diffraction is a method of broad application to other fields besides biology, such as textiles.
- The mathematical theory of diffraction from a helix proves invaluable in helping to understand the molecular structure in a fibre.

References

Astbury, W. T. (1943). *Textile Fibres Under the X-rays.* Imperial Chemical Industries Ltd/Kynoch Press, Birmingham, UK.

Cochran, W., Crick, F. H., and Vand, V. (1952). *The structure of synthetic polypeptides. I. The transform of atoms on a helix.* Acta Cryst., 5, 581–586.

Franklin, R. E. and Gosling, R. G. (1953). *Molecular configuration in sodium thymonucleate.* Nature, 171(4356), 740–741.

Fuller, W. (2019). Review of *Unravelling the double helix: the lost heroes of DNA* by Gareth Williams, Weidenfeld & Nicholson, London, 2019, 528pp., USD 23.96/16.53 (hardback/paperback) ISBN: 9781474609357. Crystallography Reviews, 25, 61–64.

Hukins, D. W. L. (1981). *X Ray Diffraction by Disordered and Ordered Systems.* Pergamon Press, Oxford, UK.

Mahendrasingam, A. et al (1986). *Time-resolved X-ray diffraction studies of the structural transition in the DNA double helix.* Science, 233, 195–197. doi:10.1126/science.372652

Rawlins, F. (1943). *Textile Fibres under the X-Rays.* Nature, 152, 90. https://doi.org/10.1038/152090e0

Watson, J. and Crick, F. (1953). *Molecular Structure of Nucleic Acids: A Structure for Deoxyribose Nucleic Acid.* Nature, 171, 737–738. https://doi.org/10.1038/171737a0

9
Powder diffraction

I have selected a biomedical example for this chapter. This chapter is mainly concerned with the chemical structure of medicines in tablet form, which are required by pharmaceutical regulators and patent approval authorities but a side topic for structural biologists. Nevertheless, we should be interested in it. Furthermore, as any single diffraction method, in this case powder diffraction, is concerned with seeking the best precision, diffraction measurements from a crushed tablet are at least studying the drug in the form in which the patient is taking it for their disease or ailment. [We note that tablets often contain other fillers but which are neutral to a patient.]

A drug-containing tablet is therefore the 'real thing' or 'real system' for research investigation. I will describe both small molecule powder diffraction and protein powder diffraction.

There are many well-documented examples of such pharmaceutically important polymorphs (see Bernstein's (2020) Chapter 10). An interesting recent case described by Sherwood et al (2022) is that of psilocybin {systematic name: 3-[2-(dimethylamino)ethyl]-1H-indol-4-yl dihydrogen phosphate} which is a zwitterionic tryptamine natural product found in numerous species of fungi known for their psychoactive properties (see Figure 9.1).

A zwitterion is a molecule that contains an equal number of positively and negatively charged functional groups. Following psilocybin's original structural elucidation and chemical synthesis in 1959, purified synthetic psilocybin

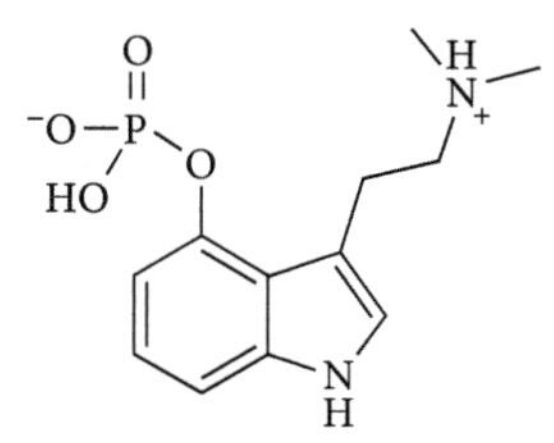

Scheme 1

Figure 9.1 The chemical structure of psilocybin.

Reproduced with the permission of the Corresponding author (Dr James Kaduk) of Sherwood et al 2022 and the IUCr Journals.

Precision and Accuracy in Biological Crystallography, Diffraction, Scattering, Microscopies, and Spectroscopies.
John R. Helliwell, Oxford University Press. © John R. Helliwell (2025). DOI: 10.1093/9780198952848.003.0009

was evaluated in clinical trials and showed promise in the treatment of various mental health disorders. This compound became the subject of a legal patent dispute (Love (2022)). A non-profit organization, Freedom to Operate, used research from chemists and crystallographers (Sherwood et al (2022)) to argue in a legal filing that a company called Compass Pathways Limited's claims that a form of synthetic psilocybin they had found was not actually an invention (Londesbrough et al (2019)). The research of Sherwood et al (2022) firmly established that there are three crystalline forms of psilocybin namely: 'Hydrate A, Polymorph A, and Polymorph B' (see Figures 9.2, 9.3, and 9.4).

There is therefore no Polymorph A′ as claimed by Compass. (Sherwood et al (2022) use the designation Polymorph A′ to distinguish it from their Polymorph A; the patent names it Polymorph A.) To challenge a patent claim required meticulous proof and that is what Sherwood et al (2022) provided. Another dimension to this legal dispute was that Compass, unlike Usona, is a for-profit company that went public in September 2020.

Let's turn to protein powder diffraction examples. Karavassili et al (2017) reviewed research findings on human insulin microcrystals exhibiting polymorphism upon physicochemical modifications (i.e. pH, ligand binding). Four new biologically active types of human insulin crystals were identified, and their structures successfully determined by a combination of powder and single crystal diffraction measurements. Such research continues for pharmaceutical products containing microcrystals with improved activity and stability for patient benefit.

The full scope of protein powder diffraction, including to larger proteins, is described in Margiolaki (2019).

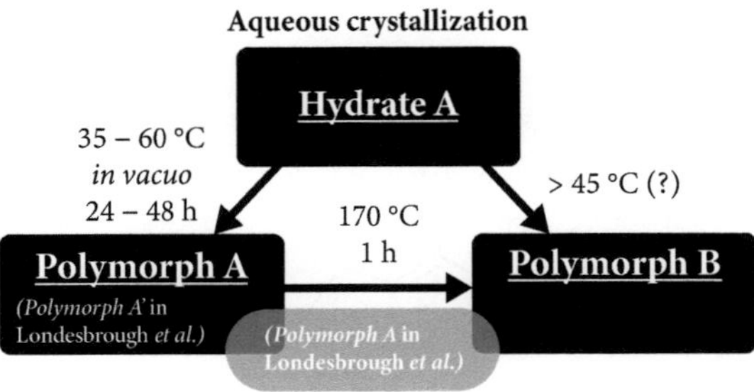

Figure 9.2 Preparation pathways for the three different chemical forms of psilocybin.

Reproduced with the permission of the Corresponding author (Dr James Kaduk) of Sherwood et al (2022) and the IUCr Journals.

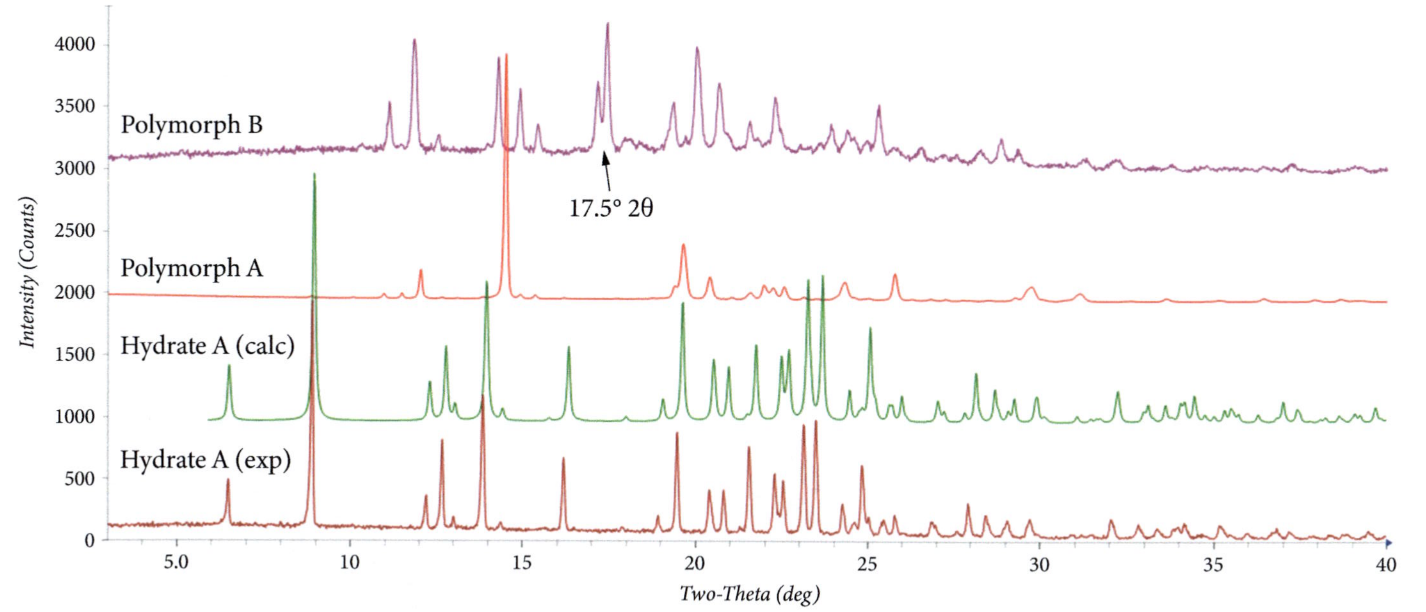

Figure 9.3 Powder X-ray diffraction patterns for polymorphs A and B, as well as the hydrated form.

Reproduced with the permission of the Corresponding author (Dr James Kaduk) of Sherwood et al (2022) and the IUCr Journals.

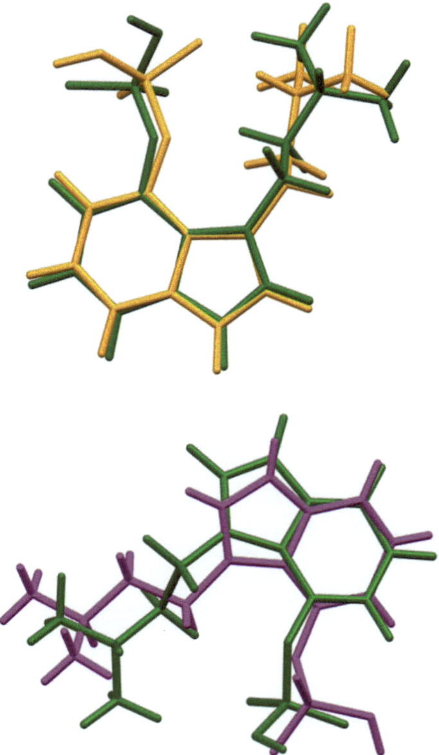

Figure 9.4 Structural superpositions of molecular structures of psilocybin. Top: the two Polymorphs A and B, and (Bottom): the Hydrate and Polymorph A.
Reproduced with the permission of the Corresponding author (Dr James Kaduk) of Sherwood et al (2022) and the IUCr Journals.

General educational materials for powder diffraction are provided at the Advanced Photon Source:—https://www.aps.anl.gov/Education/Powder-Diffraction-Educational-Materials. Home laboratory powder diffraction instrumentation details are provided by MalvernPanalytica and Bruker, each with helpful websites including application examples (see respectively https://www.malvernpanalytical.com/en/products/product-range/empyrean-range/empyrean and https://www.bruker.com/en/products-and-solutions/diffractometers-and-x-ray-microscopes/x-ray-diffractometers.html). Powder X-ray diffraction has a long history, details of which can be seen in the IUCr Teaching Pamphlet by Henry Lipson (https://www.iucr.org/education/pamphlets/16/full-text). Powder diffraction with neutrons is a large research activity across many areas of materials research (see Copley (1988)).

Key Learning Points

- Powder diffraction is essential in patenting pharmaceutical tablets as the regulatory authorities require characterization of what a patient consumes.
- Protein powder diffraction is a more specialist application; insulin, a very important medicine, provides an interesting example of this.

References

Bernstein, J. (2020). *Polymorphism in Molecular Crystals*, 2nd edition. Oxford University Press, Oxford, UK.

Copley, J. R. D. (1988). *Neutron Powder Diffraction.* NIST https://nvlpubs.nist.gov/nistpubs/Legacy/IR/nistir6204.pdf.

Karavassili, F., Valmas, A., Fili, S., Georgiou, C.D., and Margiolaki, I. (2017). *In quest for improved drugs against diabetes: The added value of X-ray powder diffraction methods.* Biomolecules, 7, 63.

Londesbrough, D. J., Brown, C., Northen, J. S., Moore, G., Patil, H. K., and Nichols, D. E. (2019). *Preparation of psilocybin, different polymorphic forms, intermediates, formulations and their use* US Patent 10519175B2, Compass Pathways Ltd, UK.

Love, S. (2022). *Inside the Dispute Over a High-Profile Psychedelic Study.* Vice Magazine https://www.vice.com/en/article/4awj3n/inside-the-dispute-over-a-high-profile-psychedelic-study.

Margiolaki, I. (2019). International Tables for Crystallography. Vol. H. *Macromolecular Powder Diffraction*, ch. 7.1, pp. 718–736.

Sherwood, A. M., Kargbo, R. B., Kaylo, K. W., Cozzi, N. V., Meisenheimer, P., and Kaduk, J. A. (2022). *Psilocybin: Crystal structure solutions enable phase analysis of prior art and recently patented examples.* Acta Cryst., C78, 36–55.

Parts of this chapter I published in 2024 in my book https://www.routledge.com/The-Scientific-Truth-the-Whole-Truth-and-Nothing-but-the-Truth/Helliwell/p/book/9781032521398

10

Small-angle solution scattering

In the pursuit of accuracy, a very interesting state of matter closer to that of the living cell is the liquid state. Solution scattering (using X-rays or neutrons) has a long and distinguished history in developing its capabilities as an analytical method. A key requirement of the method is to measure from a sufficiently dilute solution of pure macromolecules, thereby minimizing aggregation. This takes one away from the state of matter that is the living cell, which is a highly concentrated mixture of many macromolecules, and indeed highly organized in organelle compartments, membranes, and chromosomes. Nevertheless, solution scattering is a highly valued analytical method. In my own research we have used it in combination with protein crystallography and electron microscopy studies of the lobster crustacyanin (Chayen et al (2003), Rhys et al (2011)), a carotenoprotein complex.

More generally, two articles (Jacques et al (2012), Trewhella et al (2017)) describe the publication guidelines for structural modelling of small-angle scattering data from biomolecules in solution, and discuss the rationale for their application. Furthermore, the reproducibility of solution-scattering studies with X-rays or neutrons were described (Trewhella et al (2022)) based on the results of a round-robin investigation using five different standard samples of proteins. That coordinated project involved 12 SAXS (Small-Angle X-ray Scattering) and four SANS (Small-Angle Neutron Scattering) instruments around the world. The consensus scattering profiles it provided are a benchmarking set which can be used to evaluate approaches to scattering-profile prediction from atomic coordinates.

Trewhella et al's (2017) article is clearly important for anyone communicating their results and for anyone reading about them or reusing the deposited data from https://www.sasbdb.org/ or https://www.bioisis.net/. As Trewhella et al (2022) remark '*without stringent attention to data and model validation, there is significant potential for over-interpretation or even being misled*'. The authors summarize the Trewhella et al (2017) publication guidelines paper which should consider '*(i) sample quality, (ii) data acquisition and reduction, (iii) the presentation of scattering data and validation, and (iv)*

Precision and Accuracy in Biological Crystallography, Diffraction, Scattering, Microscopies, and Spectroscopies.
John R. Helliwell, Oxford University Press. DOI: 10.1093/9780198952848.003.0010

structure modelling'. They emphasize that their '*main focus remains on experiments aimed at three-dimensional structural modelling from solution SAS data. As such, SAS experiments aimed at understanding highly heterogeneous mixtures, transient species using time-resolved data, or high-throughput screening experiments are not explicitly considered as each of these important applications would have distinct attributes that need to be considered separately in detail*'.

The text above is reproduced with the permission of Professor Jill Trewhella and Dr Patrice Vachette and the IUCr Journals.

In each section, Trewhella et al (2017) provide meticulous detail on the metadata required to be made available to the reader and data reuser in their Tables 1, 2, 3, and 4. The foundation platform for the metadata at the various stages is the sasCIF (see Figure 10.1).

Obviously, the exclusion of highly heterogeneous mixtures from consideration by Trewhella (2017) et al's focus reminds us that SAXS and SANS measurements are a long way from a living cell, as remarked in the introduction to this chapter. The structure of the biological macromolecule is a model which '*is optimized such that a penalty function is minimized that includes the fit to the scattering data (i.e. χ^2, see Equation 10.1), and any other penalties*

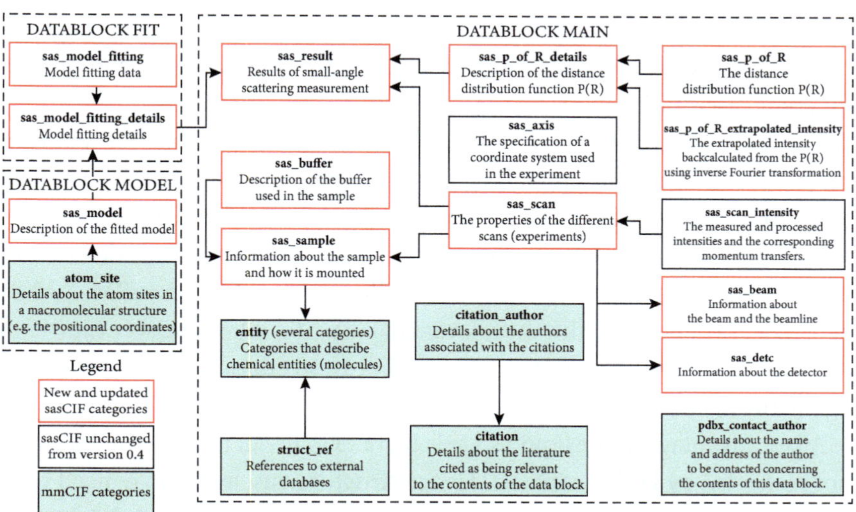

Figure 10.1 The sasCIF which was updated in 2016 (Kachala et al (2016)). The categories that remained unchanged from the previous version(s) are shown in black boxes, while updated ones are shown in red boxes. Categories from the mmCIF dictionary are in boxes with a pale-green background.
Reproduced with the permission of Dr Dmitri Svergun and the IUCr Journals.

related to restraints on the model (e.g. compactness, connectedness, distance restraints etc.)'.

$$\chi^2 = \frac{1}{N-1}\sum_{j=1}^{N}\left[\frac{I_{\text{exp}}(qj) - cI_{\text{mod}}(qj)}{\sigma(qj)}\right]^2 \qquad \text{Equation (10.1)}$$

where N is the number of points in the scattering profile, $I_{\text{exp}}(q)$ is the experimental scattering profile, $I_{\text{mod}}(q)$ is the computed scattering profile based on the three-dimensional model, c is a multiplicative scaling parameter that is used to minimize χ^2, and $\sigma(q)$ is the standard error for each measured data point. From Equation (10.1) we see that χ^2 will be smaller for data with poor statistics and conversely larger for data with vanishingly small statistical errors. Thus, while relative χ^2 values are most valuable in comparing two models against the same data set, absolute values can be less useful in comparing fits to two independent data sets.

Trewhella et al (2017) emphasize that '*Having obtained accurate and sufficiently precise data as I (q) versus q for the system of interest, provided evidence that the scattering profile is free of nonspecific aggregation or interparticle interference effects that it yields the expected M or V value, and having assessed the potential flexibility of the system, a three-dimensional modelling strategy can be selected*'. Trewhella et al (2017) conclude with detailed discussion of three different case studies.

Trewhella et al (2023) provide an update of the template tables for reporting biomolecular structural modelling of small-angle scattering data. Trewhella et al (2023) acted on the IUCr Journals' announcement for their biology journals that for any manuscript containing conventional structures determined by the most common techniques (crystallography, NMR, cryoEM, SAXS), data that are deposited with the relevant database to obtain the accession code their validation reports must be uploaded prior to editorial review. They also define the need for accuracy.

A key advantage of the liquid state is that flexibility of the macromolecule is not restricted by lattice contacts. Hence structure ensembles may need to be considered. Furthermore, multiple models may fit the measured data equally well. These and other cases, such as dynamics, time-resolved studies, and mixtures, are expanded on in the books by Svergun et al (2013) and by Lattman et al (2018), which provide many detailed examples on compact globular proteins and considerations of X-ray irradiation damage, instrumentation, and analysis.

10.1 Crustacyanin as a case study

This case study illustrates how solution scattering is often used in combination with electron microscopy and crystallography. An early electron-microscopy study of α-crustacyanin (molecular weight 320 kDa; an octamer of β-crustacyanins and 16 astaxanthins) (Zagalsky & Jones, 1982; Figure 10.2) revealed several different structures consistent in size with the value of the Stokes radius of the protein,73 Å. These included diffuse rings and both four- and five-sided structures, composed of four, six, or an indefinite number of subunits of β-crustacyanin size (40 kDa). One possible subjective interpretation was of a compact double tetrameric structure in which the two tetramers are face to face, with the subunits of one positioned in the grooves between the subunits of the other. An alternative considered was a linear array of eight β-crustacyanin molecules coiled in a helical manner into a compact configuration of between four and five subunits per turn, in which the subunits of the second turn of the helix occupy the grooves of those of the first. If viewed at different angles, such structures would provide the observed electron-micrograph patterns. The helical arrangement was supported by small-angle X-ray scattering of α- and β-crustacyanin (Dellisanti et al (2003)).

While nowadays we might immediately turn to cryoEM as a way to determine the structure of the full α-crustacyanin complex, in the past combining complementary data from electron microscopy, X-ray crystallography, and small-angle X-ray scattering was used to unravel such a case. The molecular weight of the complex, at 320 kDa, also challenged these methods to determine its structural arrangement. We initiated SAXS experiments, including both the small- and the wider-angle region, and applied symmetry constraints from the electron micrographs. In addition, as mentioned above, Dellisanti et al (2003) also investigated the α-crustacyanin complex with SAXS, which also used the electron-micrograph constraints. In their case they assumed a possible helical layout proposed as an option by Zagalsky & Jones (1982). Dellisanti et al (2003) presented evidence of the α-crustacyanin being of cylindrical shape with axial length 238 Å and a radius of 47.5 Å, in a helical arrangement, pointing out that '*the complexity of this non-globular scaffolding containing many discontinuities as well as the relatively large size of the protein (2840 residues) may be the reasons why the classical ab initio shape restoration programmes have, unfortunately, in this case been exceeded*'. We also harnessed the electron-micrograph information of Zagalsky & Jones (1982), but with the option of a fourfold look to the electron micrographs and thereby applied that symmetry constraint instead. Grossmann and coworkers' SAXS measurements on

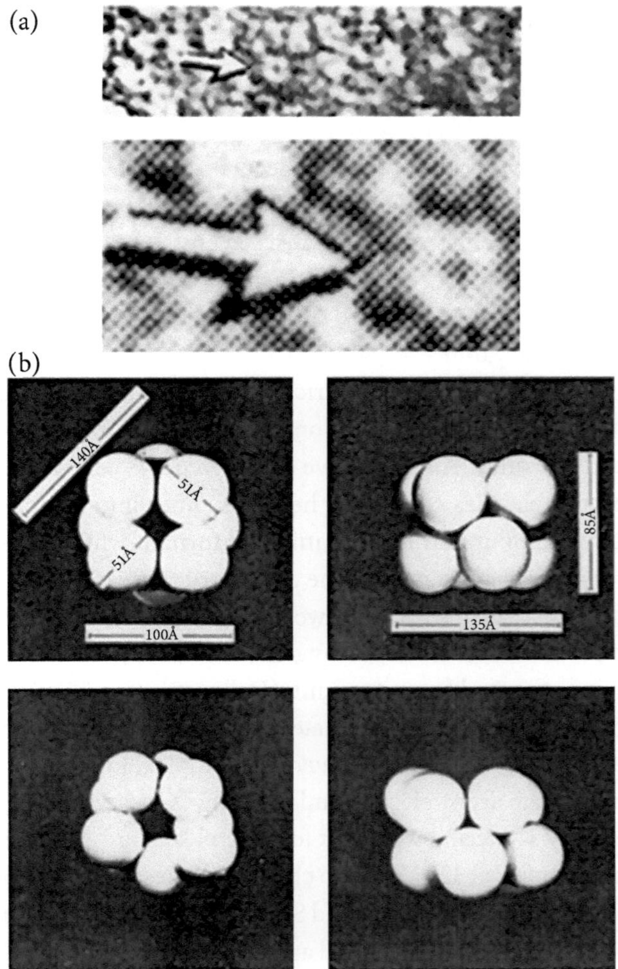

Figure 10.2 (*a*) Electron micrographs of α-crustacyanin. (*b*) Proposed schematic model for the quaternary structure of α-crustacyanin. Each sphere represents a β-crustacyanin. Top, double tetramer; bottom, helical coil.
Reprinted from Zagalsky & Jones (1982), copyright (1982) with permission from Elsevier.

SRS station 2.1 (Figure 10.3) concluded with a plausible 'piano-stool' layout interpretation. This had approximate dimensions 80 Å across the seat of the stool, 150 Å in height, and 130 Å width between the splayed legs of the stool. The overall diagonal maximum dimension would be 220 Å. This was at variance with the option that Dellisanti et al (2003) had favoured. Even though our shape restorations from the scattering data consistently produced comparable versions of the piano-stool arrangement, we too concluded that improvements

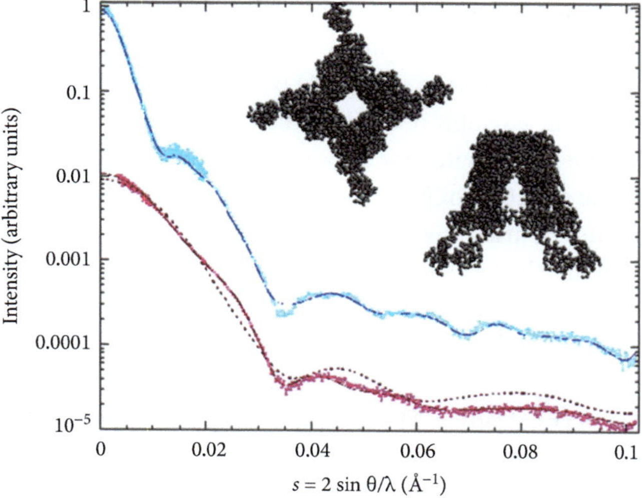

Figure 10.3 Small-angle X-ray scattering data measured at the Daresbury SRS station 2.1 from solutions of α-crustacyanin (top curve, with error bars) and β-crustacyanin (bottom curve, also with error bars) as a function of the momentum transfer s (which is defined by the X-ray wavelength λ and the scattering angle 2θ). The curves have been displaced for better visualization. The smooth curve fitting the α-crustacyanin scattering profile results from a shape restoration with fourfold molecular symmetry. The inset depicts a representative shape model in two orthogonal orientations highlighting the proposed 'piano-stool' layout. Each amino acid of full-length crustacyanin is represented by a sphere. In contrast, the red curve superimposed on the β-crustacyanin profile results from a scattering pattern simulation using the β-crustacyanin crystal structure of Cianci et al (2002). As expected, the dotted curve representing the simulation based on the A1 apocrustacyanin homodimer of Cianci et al (2001) clearly provides a very poor fit to the experimental data.
From Chayen et al (2003) with the permission of IUCr Journals.

in cryoEM offered the way to better define a molecular complex to assist the SAXS data interpretation.

Why was it so interesting to get the detailed layout of α-crustacyanin? These structural details were sought to explain the final 50 nm bathochromic shift stemming from the complexation of the eight β-crustacyanins. Cianci et al's (2002) X-ray β-crustacyanin crystal structure publication had commented on different kinds of final plausible twistings of the bound astaxanthin conformations as a result of complexation of eight β-crustacyanins into the fully assembled α-crustacyanin.

The α-crustacyanin was further investigated by Rhys et al (2011). A low-resolution structure of α-crustacyanin was determined to 30 Å resolution using negative-stain electron microscopy (EM) with single-particle averaging. The protein, which is an assembly of eight β-crustacyanin dimers, was asymmetrical and rather open in layout. A model was built to the EM map using the X-ray crystallographic structure of β-crustacyanin guided by PISA interface analyses (Krissinel and Henrick (2007)). The model had a theoretical sedimentation coefficient that matched well with the experimentally derived value from sedimentation velocity analytical ultracentrifugation. Additionally, the EM model had similarities to models calculated independently by rigid-body modelling to small-angle X-ray scattering (SAXS) data and extracted *in silico* from the β-crustacyanin crystal lattice. Theoretical X-ray scattering from each of those models was in reasonable agreement with the experimental SAXS data and together suggested an overall design for the α-crustacyanin assembly.

Cedri et al (2025) now have a cryoEM structure of α-crustacyanin purified from the North American lobster shell.

To conclude this short description of the β-crustacyanin studies, let's come back to precision and accuracy. The β-crustacyanin crystal structure at 3.2 Å diffraction resolution is not the most precise structure one can hope for, versus a higher diffraction resolution, but was sufficient for Cianci et al (2002) to itemize the atomic details of the astaxanthin bathochromic shift of 100 nm. A challenging portion of the β-crustacyanin X-ray crystal structure determination was placement of two key water molecules. Inevitably for X-rays at 3.2 Å diffraction resolution, the protonation states of two key histidines involved in the complexation of the two astaxanthins in the β-crustacyanin complex were not determined. Later, we investigated the effectiveness of protonation state prediction software on all ionizable amino acid types (Asp, Glu, His, Lys, Arg) finding it ineffective for histidines (Fisher et al (2009)). Thus, the determination of these two β-crustacyanin histidine protonation states even by prediction remains unknown. However, a more positive aspect was that the colour of the β-crustacyanin crystals was blue at both room temperature and at 100 K, and the room temperature colour of the β-crustacyanin in solution was likewise blue. Furthermore, the SAXS data obtained from β-crustacyanin in solution, unlike the apo dimer, matched that predicted from the X-ray crystal structure of Cianci et al (2002) (see Figure 10.3). We can conclude, then, that we gained an accurate picture of the molecular basis of the coloration mechanism of β-crustacyanin by using complementary diffraction methods of X-ray crystallography and SAXS, with the colours of the live lobster providing additional evidence.

Key Learning Points

- Solution small-angle scattering as an analytical method has a long history.
- It is widely applied.
- X-rays or neutrons can be used (SAXS and SANS) thus exploiting their complementary characteristics as scattering probes.
- A solution small-angle scattering database is available and is well organized by the community.

References

Chayen, N. E., Cianci, M., Grossmann, J. G., Habash, J., Helliwell, J. R., Nneji, G. A., Raftery, J., Rizkallah, P. J., and Zagalsky, P. F. (2003). *Unravelling the structural chemistry of the colouration mechanism in lobster shell.* Acta Cryst., D59, 2072–2082.

Cianci, M., Rizkallah, P. J., Olczak, A., Raftery, J., Chayen, N. E., Zagalsky, P. F., and Helliwell, J. R. (2001). *Structure of lobster apocrustacyanin A1 using softer X-rays.* Acta Cryst., D57, 1219–1229.

Cianci, M., Rizkallah, P. J., Olczak, A., Raftery, J., Chayen, N. E., Zagalsky, P. F., and Helliwell, J. R. (2002). *The molecular basis of the coloration mechanism in lobster shell: β-Crustacyanin at 3.2-Å resolution.* Proc. Natl Acad. Sci. USA, **99**, 9795–9800.

Cedri, M. C., Bansia, H., Amici, A. T., Moretti, P., Ortore, M. G., McCarthy, A., Mueller-Dieckmann, C., Lingas, R., Durbeej, B., Raffaelli, N., Wang, T., des Georges, A., and Cianci, M. (2025) *The 2.75 Å cryo-EM structure of the blue pigment a-crustacyanin from American lobster reveals an HPR protein for interaction with the carotenoid astaxanthin.* Submitted.

Dellisanti, C. D., Spinelli, S., Cambillau, C., Findlay, J. B. C., Zagalsky, P. F., Finet, S., and Receveur-Brechot, V. (2003). *Quaternary structure of alpha-crustacyanin from lobster as seen by small-angle X-ray scattering.* FEBS Lett., 544, 189–193.

Fisher, S. J., Wilkinson, J., Henchman, R. H., and Helliwell, J. R. (2009). *An evaluation review of the prediction of protonation states in proteins versus crystallographic experiment.* Crystallography Reviews, 15 (4), 231–259. https://doi.org/10.1080/08893110903213700

Jacques, D. A., Guss, J. M., Svergun, D. I., and Trewhella, J. (2012). *Publication guidelines for structural modelling of small-angle scattering data from biomolecules in solution.* Acta Cryst., D68, 620–626.

Kachala, M., Westbrook, J., and Svergun, D. (2016). *Extension of the sasCIF format and its applications for data processing and deposition.* J. Appl. Cryst., 49, 302–310.

Krissinel, E. and Henrick, K. (2007). *Inference of macromolecular assemblies from crystalline state.* J. Mol. Biol., 372, 774–797.

Lattman, E. E., Grant, T. D., and Snell E. H. (2018). *Biological Small Angle Scattering Theory and Practice.* IUCr Book Series. Oxford University Press, Oxford, UK.

Rhys, N. H., Wang, M.-C., Jowitt, T. A., Helliwell, J. R., Grossmann, J. G., and Baldock, C. (2011). *Deriving the ultrastructure of α-crustacyanin using lower-resolution structural and biophysical methods.* J. Synchrotron Rad., 18, 79–83.

Svergun, D. I., Koch, M. H. J., Timmins, P. A., and May, R.P. (2013). *Small Angle X-Ray and Neutron Scattering from Solutions of Biological Macromolecules.* IUCr Texts on Crystallography, No. 19. Oxford University Press, Oxford, UK.

Trewhella, J., Duff, A. P., Durand, D., Gabel, F., Guss, J. M., Hendrickson, W. A., Hura, G. L., Jacques, D. A., Kirby, N. M., Kwan, A. H., Perez, J., Pollack, L., Ryan, T. M., Sali, A., Schneidman-Duhovny, D., Schwede, T., Svergun, D. I., Sugiyama, M., Tainer, J. A., Vachette, P., Westbrook, J., and Whitten, A. E. (2017). *2017 publication guidelines for structural modelling of small-angle scattering data from biomolecules in solution: An update.* Acta Cryst., D73, 710–728.

Trewhella, J., Vachette, P., Bierma, J., Blanchet, C., Brookes, E., Chakravarthy, S., Chatzimagas, L., Cleveland, T. E., Cowieson, N., Crossett, B., Duff, A. P., Franke, D., Gabel, F., Gillilan, R. E., Graewert, M., Grishaev, A., Guss, J. M., Hammel, M., Hopkins, J., Huang, Q., Hub, J. S., Hura, G. L., Irving, T. C., Jeffries, C. M., Jeong, C., Kirby, N., Krueger, S., Martel, A., Matsui, T., Li, N., Perez, J., Porcar, L., Prange, T., Rajkovic, I., Rocco, M., Rosenberg, D. J., Ryan, T. M., Seifert, S., Sekiguchi, H., Svergun, D., Teixeira, S., Thureau, A., Weiss, T. M., Whitten, A. E., Wood, K., and Zuo, X. (2022). *A round-robin approach provides a detailed assessment of biomolecular small-angle scattering data reproducibility and yields consensus curves for benchmarking.* Acta Cryst., D78, 1315–1336.

Trewhella, J., Jeffries, C. M., and Whitten, A. E. (2023). *2023 update of template tables for reporting biomolecular structural modelling of small-angle scattering data.* Acta Cryst., D79, 122–132.

Zagalsky, P. F. and Jones, R. (1982). *Quaternary structures of the astaxanthin-proteins of* Velella velella, *and of α-crustacyanin of lobster carapace, as revealed in electron microscopy.* Comp. Biochem. Physiol., 71, 237–242.

11
Electron microscopy (EM)

EM has made extensive strides in its capabilities for structural biology in the past decade. This was recognized with the Nobel Prize for Chemistry awarded to Jacques Dubochet, Joachim Frank, and Richard Henderson in 2017. The headline citation stated that the Prize was awarded '*for developing cryo-electron microscopy for the high-resolution structure determination of biomolecules in solution*'. A short résumé of the achievements of each Nobel Laureate is available at the Nobel Prize 2017 website https://www.nobelprize.org/prizes/chemistry/2017/summary/.

In the mid-1970s I was undertaking my PhD at Oxford University Laboratory of Molecular Biophysics, and I vividly recall Richard Henderson's lecture to us all. I remember that he emphasized the importance of adhering to as low a dose of electrons to the specimen as possible.

Skipping forward to the 2020s, the PDB convened a workshop involving 47 experts to determine the community recommendations on cryoEM data archiving and validation (Kleywegt et al (2024)).

It is first worth recalling that in 2002, another community initiative led to the establishment of the Electron Microscopy Data Bank (EMDB) (Tagari et al (2002)). EMDB archives contain processed experimental data from a variety of 3D EM modalities, most notably single-particle analysis (SPA), electron tomography, sub-tomogram averaging (STA), helical reconstruction (HR) and electron crystallography (EC) (wwPDB consortium (2024)). Furthermore, later the Electron Microscopy Public Image Archive (EMPIAR) for raw 2D image data underpinning 3DEM volumes deposited in EMDB was established in 2013 (for the most recent details see Ludin et al (2023)).

In the past decade there has been a very big increase in single particle cryoEM structures, which has necessitated a review of parameters of structure quality. This has been led by the PDB (Kleywegt et al (2024)). The new methodology has brought great advantages for structural biology as it has made tractable structural studies of proteins (e.g. membrane

Precision and Accuracy in Biological Crystallography, Diffraction, Scattering, Microscopies, and Spectroscopies.
John R. Helliwell, Oxford University Press. © John R. Helliwell (2025). DOI: 10.1093/9780198952848.003.0011

proteins) and large complexes that failed to crystallize, or if they did, diffracted poorly.

As in crystallography, cryoEM structures are fitted to maps (electrostatic potential maps). However they are refined 'only' in real space (such as in Coot (Emsley et al (2010)) or e.g. in use of Isolde (Croll (2018)) which has dynamics incorporated. In crystallography, model refinement against the diffraction data leads to improved electron density maps. CryoEM map modification, and ease of interpretation, is evolving; for a recent description of procedures used see sections 6 (Map sharpening) and 7 (Model building and fitting) of Pintilie & Chiu (2021).

In evaluating cryoEM structures, Kleywegt et al (2024) neatly provide a resumé of the details provided in their PDB Validation Reports. These are invaluable in evaluating the structures of a particular theme, as I did when writing a topical review of the enzyme family glucose-6-phosphate dehydrogenase (G6PDH) and its 3D structures from crystallography and electron cryo-microscopy (Hanau and Helliwell (2024)); see especially their Table 2 as well as Table 1. That said, I greatly appreciated the guidance of Dr Tom Burnley of the UK CCPEM software project based at Harwell Campus, UK to navigate my way through the cryoEM Validation Reports. My scrutiny of the cryoEM maps was straightforward using the excellent molecular graphics program Coot (Emsley et al (2010)). I was impressed by the quality of these cryoEM electrostatic potential maps of the G6PDH enzyme structures.

Kleywegt et al (2024) described the aims of their 2020 PDB meeting as follows. '*(1) To provide advice on how to improve (meta)data deposition in the PDB and EMDB. (2) To review the contemporaneous, preliminary 3DEM validation reports and obtain feedback and suggestions for their improvement. (3) To discuss potential additional model-only, map-only and map–model validation metrics for single-particle cryoEM depositions*'.

Kleywegt et al (2024) stated 10 recommendations which '*mostly involved improving the wwPDB deposition system OneDep, thereby improving the data content of both the PDB and EMDB archives to facilitate wider and better use of these data*'. There were additional recommendations in specific categories of models and maps.

Kleywegt et al (2024) made clear that '*"bleeding edge" metrics are avoided until more experience has been gained with them and their applicability, performance, utility and limitations are better understood*'.

The text above quoted from Kleywegt et al (2024) is with the permission of Dr Gerard Kleywegt and the IUCr Journals.

11.1 Case study: A computational protocol to describe coordinate uncertainty and the density at each atom in cryoEM

Hryc et al (2017) is a highly significant proposal not least as it is consistent with efforts initiated by Cruickshank (1999) for introducing coordinate error estimates in protein crystallography. The criterion normally used though in cryoEM is to evaluate cross-correlation of the independent/random 3D volumes in an electrostatic potential map via the 'gold standard Fourier shell coefficient (GSFSC)' (Heel and Schatz (2005)). This is used to define the nominal resolution—namely when the FCS is 0.143. This metric with examples has been recently reviewed by Pintilie & Chiu (2021).

Hryc et al (2017) describe their initiative as follows:

> *Significance: Electron cryomicroscopy is a rapidly growing field for macromolecular structure determination. We establish a computational protocol to construct a de novo atomic model from a cryo-EM density map, along with associated metadata that describe coordinate uncertainty and the density at each atom.*

To obtain their coordinate uncertainty estimates they generated two independent models, optimized for two half-maps. These provided an understanding of the level of uncertainty within the whole map. Both half-data sets (~3.4 Å in resolution) were modelled independently. The resulting root-mean-square deviation (RMSD) of 0.56 Å between the Cα-atom positions of the two independent model sets indicated the reproducibility of the data. Also, these two independent models agreed well with their model built using the whole data set (i.e. ~ 0.6 Å for both data sets).

Let's now assess a very recent case study.

11.2 Case study of inhibition of *M. tuberculosis* and human ATP synthase by BDQ and TBAJ-587

This study (Zhang et al (2024)) really does look important:

> *Our study provides a better understanding of the similarities and differences in bedaquiline (BDQ)'s modes of binding to the human versus the* M. tuberculosis *ATP synthase and presents an opportunity for the design of derivatives of BDQ as new TB drugs with decreased side effects and a better ability to stave off resistance.*

This study was underpinned by the following cryoEM structures (Table 11.1). Figure 2d of Zhang et al (2024) is reproduced here as Figure 11.1.

Since Kleywegt et al (2024) do not endorse the approach of Hryc et al (2017) for securing coordinate uncertainty estimates, or offer an alternative,

Table 11.1 CryoEM structures underpinning Zhang et al (2024).

PDB Code	Description	Map resolution estimate of the depositors
8J0T	CryoEM structure of Mycobacterium tuberculosis ATP synthase in the apo-form	2.8Å
8J58	CryoEM structure of Mycobacterium tuberculosis ATP synthase Fo in the apo-form	3.15Å
8J0S	CryoEM structure of Mycobacterium tuberculosis ATP synthase in complex with bedaquiline(BDQ)	2.58Å
8J57	CryoEM structure of Mycobacterium tuberculosis ATP synthase Fo in complex with bedaquiline(BDQ)	2.85Å
8JR0	CryoEM structure of Mycobacterium tuberculosis ATP synthase in complex with TBAJ-587	2.80Å
8JR1	CryoEM structure of Mycobacterium tuberculosis ATP synthase Fo in complex with TBAJ-587	3.17Å
8KI3	EM Structure of the human ATP synthase bound to bedaquiline (composite)	2.89Å
8KHF	EM Structure of the human ATP synthase bound to bedaquiline (membrane domain)	3.13Å
8H9S	EM Human ATP synthase state 1 (combined)	2.53Å
8H9F	EM Human ATP synthase state 1 subregion 3	2.69Å

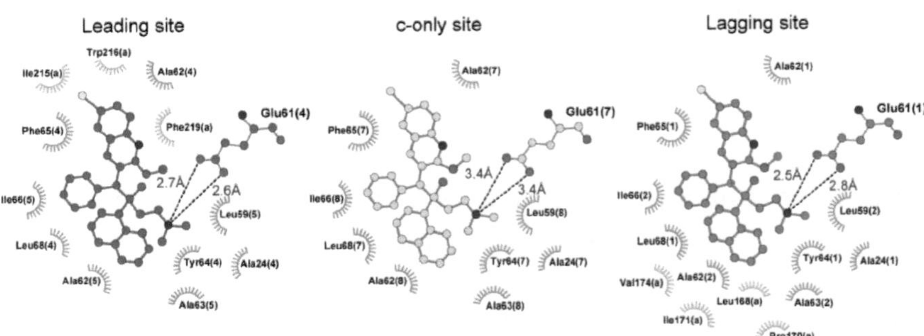

Figure 11.1 Interactions between *M. tuberculosis* ATP synthase and bedaquiline (BDQ). Their Figure 2d caption states '*Two-dimensional plot of the interactions between BDQ and M. tuberculosis ATP synthase. Hydrogen bonds are represented by black dashed lines*'.

Reproduced from Zhang, Y., Lai, Y., Zhou, S., et al (2024). *Inhibition of M. tuberculosis and human ATP synthase by BDQ and TBAJ-587.* Nature, 631, 409–414 with the permission of Prof. Hongri Gong and Springer Nature.

the Zhang et al (2024) marked distances in Figure 11.1 appear to have no clear foundation, as they have no uncertainty estimates. This is just one example in *Nature* in recent years that feature such interaction distances without uncertainty estimates. Note that these are not covalently bonded atoms, for which dictionary restraints apply, they are non-covalent, and therefore putative, interactions.

Key Learning Points

- CryoEM has developed into a high-resolution method.
- Validation guidelines from the PDB are extensive.
- Coordinate error estimation is feasible from half-maps' fitting and comparisons of the respective models, but not yet endorsed by the PDB.
- CryoEM non-covalent distances shown in publications without error estimates must be considered more critically than currently.

References

Croll, T. I. (2018). *ISOLDE: A physically realistic environment for model building into low-resolution electron-density maps.* Acta Cryst., D74, 519–530.

Cruickshank, D. W. J. (1999). *Remarks about protein structure precision.* Acta Cryst., D55, 583–601.

Emsley, P., Lohkamp, B., Scott, W. G., and Cowtan, K. (2010). *Features and development of Coot.* Acta Cryst., D66, 486–501.

Hanau, S. and Helliwell, J. R. (2024). *Glucose-6-phosphate dehydrogenase and its 3D structures from crystallography and electron cryo-microscopy.* Acta Cryst., F80, 236–251.

Hryc, C.F., Chen, D., Afonine, P. V., Jakana, J., Wang, Z., Haase-Pettingell, C., Jiang, W., Adams, P. D., King, J. A., Schmid, M. F., and Chiu, W. (2017). *Accurate model annotation of a near-atomic resolution cryo-EM map.* Proc. Natl. Acad. Sci. U.S.A., 114 (12) 3103–3108. https://doi.org/10.1073/pnas.1621152114

Heel, M. van and Schatz, M. (2005). *Fourier shell correlation threshold criteria.* J. Struct. Biol., **151**, 250–262.

Ludin, A., Korir, P. K., Somasundharam, S., Weyand, S., Cattavitello, C., Fonseca, N., Salih, O., Kleywegt, G. J., and Patwardhan, A. (2023). *EMPIAR: the Electron Microscopy Public Image Archive.* Nucleic Acids Res., 51 (D1), D1503–D1511. doi: 10.1093/nar/gkac1062. Erratum in: Nucleic Acids Res., 51 (7), 3499. doi: 10.1093/nar/gkad201. PMID: 36440762; PMCID: PMC9825465.

Kleywegt, G. J., Adams, P. D., Butcher, S. J., Lawson, C. L., Rohou, A., Rosenthal, P. B., Subramaniam, S., Topf, M., Abbott, S., Baldwin, P. R., Berrisford, J. M., Bricogne, G., Choudhary, P., Croll, T. I., Danev, R., Ganesan, S. J., Grant, T., Gutmanas, A., Henderson, R., Heymann, J. B., Huiskonen, J. T., Istrate, A., Kato, T., Lander, G. C., Lok, S.-M., Ludtke, S. J., Murshudov, G. N., Pye, R., Pintilie, G. D., Richardson, J. S., Sachse, C., Salih, O.,

Scheres, S. H. W., Schroeder, G. F., Sorzano, C. O. S., Stagg, S. M., Wang, Z., Warshamanage, R., Westbrook, J. D., Winn, M. D., Young, J. Y., Burley, S. K., Hoch, J. C., Kurisu, G., Morris, K., Patwardhan, A., and Velankar, S. (2024). *Community recommendations on cryoEM data archiving and validation.* IUCrJ, 11, 140–151.

Pintilie, G. and Chiu, W. (2021). *Validation, analysis and annotation of cryo-EM structures.* Acta Cryst., D77, 1142–1152.

Tagari, M., Newman, R., Chagoyen, M., Carazo, J. M., and Henrick, K. (2002). *New electron microscopy database and deposition system.* Trends Biochem Sci., 27 (11), 589. doi: 10.1016/s0968-0004(02)02176-x. PMID: 12417136

wwPDB consortium (2024). *EMDB—the Electron Microscopy Data Bank.* Nucleic Acids Res., **52**, D456–D465.

Zhang, Y., Lai, Y., Zhou, S., et al (2024). *Inhibition of M. tuberculosis and human ATP synthase by BDQ and TBAJ-587.* Nature, 631, 409–414. https://doi.org/10.1038/s41586-024-07605-8

12
X-ray absorption spectroscopy (XAS)

X-ray absorption spectroscopy involves the pre-edge, on-edge, and post-edge extended X-ray absorption fine structure (EXAFS) regions when a tuneable X-ray beam illuminates a sample. The oxidation state of an absorbing atom affects the precise photon energy of the absorption edge because the to-be-ejected bound electron is more strongly held by the protons in the nucleus of the ion than the neutral atom. The EXAFS data can provide precise information on the distances and type of nearest neighbours. These distances are more precise than those usually derived from macromolecular crystallography.

There are situations where the information gained from XAS is very valuable, such as a relatively low X-ray dose. The use of XAS to characterize the photosystem II Mn_4CaO_5 oxygen-evolving complex (OEC) with a lower dose of X-rays than needed in a macromolecular crystallography data collection proved very valuable (Yano et al (2005)). Not only did it allow precise characterization of the manganese ion oxidation states (II, III, or IV) in the four so-called S states known as the 'Kok cycle' (Joliot et al (1969), Kok et al (1970)), it also impacted on researchers devising new ways of measuring the macromolecular X-ray crystallography data. These experiments are described in Case study 12.1.

12.1 Case study: Photosystem II

The crystallographic 3D analyses of the photosystem II (PSII) multi-macromolecular complex at the heart of photosynthesis has been one of the outstanding structural biochemistry challenges of the modern era. The role of synchrotron X-radiation and specifically the third-generation synchrotrons ESRF and SPRing8 macromolecular crystallography beamlines have been paramount. The overall 3D architecture and its role was described by Ferreira et al (2004), Grotjohann et al (2004) and Loll et al (2005). In addition, the use of the Stanford Synchrotron Radiation Laboratory (SSRL) has provided important guidance from the EXAFS/XANES measurements of the X-ray dosage tolerance of the $CaMn_4$ oxygen-evolving complex (OEC), and where the absorbed X-rays liberating electron free radicals upset the redox

Precision and Accuracy in Biological Crystallography, Diffraction, Scattering, Microscopies, and Spectroscopies. John R. Helliwell, Oxford University Press. © John R. Helliwell (2025). DOI: 10.1093/9780198952848.003.0012

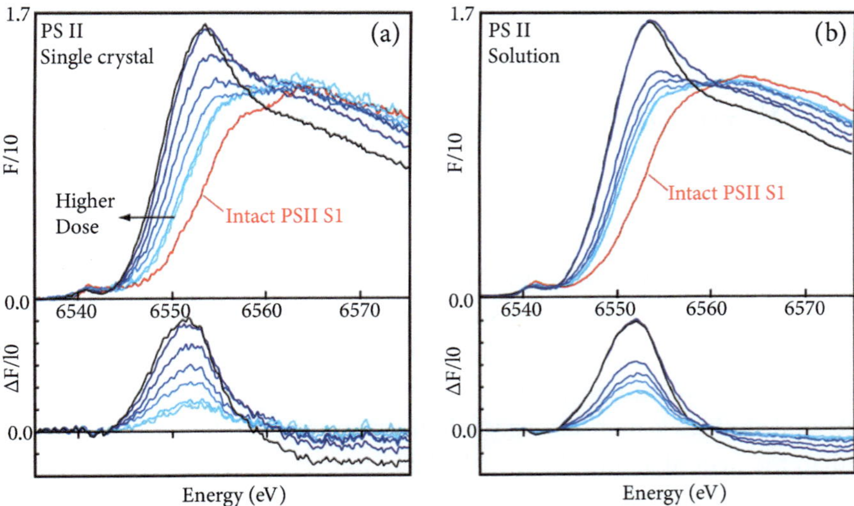

Figure 12.1 Evidence for X-ray radiation damage to PS II during EXAFS/XANES. Mn XANES of PS II versus X-ray dose and XANES of inorganic model compounds. (*A*) Mn K edge shift of PS II crystals as a function of X-ray dose at 13.3 keV (0.933 Å) and 100 K (*Upper*) and the difference spectra (*Lower*). The spectrum at the highest inflection point energy is from an undamaged PS II crystal. The X-ray exposures were 0.14, 0.21, 0.25, 0.54, 0.95, 2.3, and 5.0 $\times$ 10^{10} photons per micron^2 (light-blue to black lines). An average dose of ~ 3.5 $\times$ 10^{10} photons per micron^2 was used for X-ray diffraction studies. Exposure was at 100 K, and all XANES were collected at 10 K at low dose (1 $\times$ 10^7 per micron^2). XANES and difference spectra showed an increase in amplitude at ~6552 eV, providing definite evidence that exposure to X-rays caused photoreduction to Mn(II) in PS II crystals from Mn(III2,IV2). At a dose of 2.3 $\times$ 10^{10} photons per micron^2, equal to 66% of the average dose used for diffraction measurements, 80% of the Mn in PS II is present as Mn(II). (*B*) A similar trend in the XANES (*Upper*) and the difference spectra (*Lower*) was seen for PS II solutions. The X-ray doses used for exposure were 0.03, 0.05, 0.10, 0.17, 0.32, 0.76, and 1.4 $\times$ 10^{10} photons per micron^2 (light-blue to black lines). At a dose of 1.4 $\times$ 10^{10} per micron^2, equal to 40% of the average dose used for diffraction measurements, 90% of the Mn in PS II was present as Mn(II).

From: J. Yano, J. Kern, K.-D. Irrgang, M. J. Latimer, U. Bergmann, P. Glatzel, Y. Pushkar, J. Biesiadka, B. Loll, K. Sauer, J. Messinger, A. Zouni, and V. K. Yachandra (2005). *X-ray damage to the Mn_4Ca complex in single crystals of photosystem II: A case study for metalloprotein crystallography.* PNAS, 102, 12047–12052. Reproduced with permission of the authors and Copyright (2005) National Academy of Sciences, U.S.A.

state of the manganese ions and thereby the integrity of the OEC metal atom cluster (Figure 12.1; from Yano et al (2005); see also Robblee et al (2001)). EXAFS (extended X-ray absorption fine structure) is sensitive to

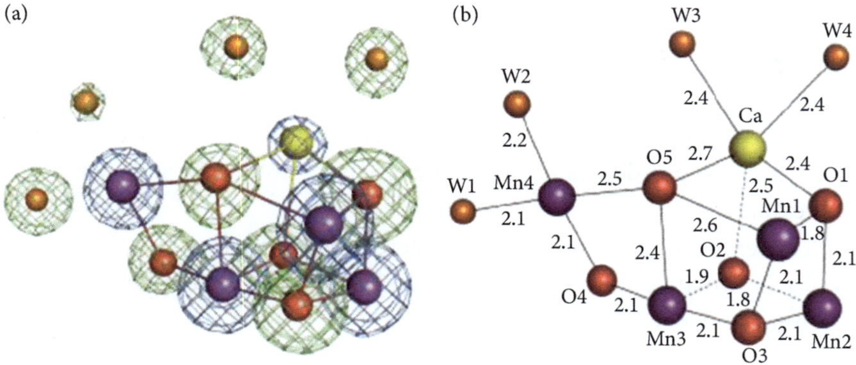

Figure 12.2 Structure of the Mn_4CaO_5 cluster. a, Determination of individual atoms associated with the Mn_4CaO_5 cluster. The structure of the cluster was superimposed with the $2F_o$-F_c map (blue) contoured at 5σ for manganese and calcium atoms, and with the omit map (green) contoured at 7σ for oxygen atoms and water molecules. b, Interatomic distances (in Å) between metal atoms and oxo bridges or water molecules.

From: Y. Umena, K. Kawakami, J.-R. Shen, and N. Kamiya (2011). *Crystal structure of oxygen-evolving photosystem II at a resolution of 1.9 Å*. Nature, 473, 55–60. Reproduced with permission of the authors and Springer Nature.

metal–metal ligand distances to a precision of 0.02 Å. The macromolecular crystallography highest resolution study of the PSII was at 1.9 Å diffraction resolution and revealed the whole OEC intact (Figure 12.2; Umena et al (2011). [Note: The structure and function of PSII are studied by a large number of research laboratories worldwide and there are many important contributions.] The breakthrough was the use of a new 'helical X-ray diffraction data scan', with the X-ray beam tracing fresh portions of the needle-shaped PSII crystal throughout data collection, to better control the X-ray dose. That study revealed the bond distances and angles in the OEC but also examined their standard uncertainties using the diffraction precision index, which for an 'average atom' was 0.11 Å (Umena et al (2011)). Thus, the precise manganese ion redox changes as a function of the S-states photo-cycle converting carbon dioxide into oxygen (Joliot et al (1969)) were put into a 3D structural framework (see eg Pushkar et al (2008)). This set the platform for future X-ray crystallography and X-ray spectroscopic studies of the structural intermediates (S states). In parallel with these basic science studies, the potential application of the OEC in materials science mimics of PSII could have far-reaching implications for fuel cells and thereby energy conservation in a move away from fossil fuels (see Birch (2009) and linked letter from Barber (2009)).

Key Learning Points

- X-ray absorption spectroscopy (XAS) can be used to study metal centres in crystals, powders, amorphous materials, or solutions, and so is very flexible.
- A lower X-ray dose is needed than macromolecular X-ray crystallography.
- An informative case study of recent times is of the oxygen-evolving complex in photosystem II.

References

Barber, J. (2009). *Letters: Artificial leaf omission.* https://www.chemistryworld.com/opinion/letters-june-2009/3005443.article

Birch, H. (2009). *The artificial leaf: Using sunlight to split water molecules and form hydrogen fuel is one of the most promising tactics for kicking our carbon habit.* http://www.rsc.org/images/Energy_ChemistryWorldMay09_tcm18-151116.pdf

Ferreira, K. N., Iverson, T. M., Maghlaoui, K., Barber, J., and Iwata, S. (2004). *Architecture of the photosynthetic oxygen-evolving center.* Science, 303, 1831–1838.

Grotjohann, I., Jolley, C., and Fromme, P. (2004). *Evolution of photosynthesis and oxygen evolution: Implications from the structural comparison of Photosystems I and II.* Physical Chemistry Chemical Physics, 6(20), 4743–4753.

Joliot, P., Barbieri, G., and Chabaud, R. (1969). *Un nouveau modele des centres photochimiques du systeme II.* Photochem. Photobiol., 10, 309–329.

Kok, B., Forbush, B., and McGloin, M. (1970). *Cooperation of charges in photosynthetic O_2 evolution-I. A linear four step mechanism.* Photochemistry and Photobiology, 11(6), 457–475.

Loll, B., Kern, J., Saenger, W., Zouni, A., and Biesiadka, J. (2005). *Towards complete cofactor arrangement in the 3.0 Å resolution structure of photosystem II.* Nature, 438, 1040–1044.

Pushkar, Y., Yano, J., Sauer, K., Boussac, A., and Yachandra, V. K. (2008). *Structural changes in the Mn_4Ca cluster and the mechanism of photosynthetic water splitting.* PNAS, 105, 1879–1884.

Robblee, J. H., Cince, R. M., and Yachandra, V. K. (2001). *X-ray spectroscopy-based structure of the Mn cluster and mechanism of photosynthetic oxygen evolution.* Biochim. Biophys. Acta, 1503, 7–23.

Umena, Y., Kawakami, K., Shen, J.-R., and Kamiya, N. (2011). *Crystal structure of oxygen-evolving photosystem II at a resolution of 1.9 Å.* Nature, 473, 55–60.

Yano, J., Kern, J., Irrgang, K.-D., Latimer, M. J., Bergmann, U., Glatzel, P., Pushkar, Y., Biesiadka, J., Loll, B., Sauer, K., Messinger, J., Zouni, A., and Yachandra, V. K. (2005). *X-ray damage to the Mn_4Ca complex in single crystals of photosystem II: A case study for metalloprotein crystallography.* PNAS, 102, 12047–12052.

13
NMR

Nuclear magnetic resonance (NMR) of biological macromolecules in solution and in the solid state are important methods for structure determination. The 2002 Nobel Prize for Chemistry to Kurt Wüthrich '*for his development of nuclear magnetic resonance spectroscopy for determining the three-dimensional structure of biological macromolecules in solution*' is an indicator of its importance. [This prize was shared, with half jointly to John B. Fenn and Koichi Tanaka '*for their development of soft desorption ionization methods for mass spectrometric analyses of biological macromolecules*'.]

Solid-state NMR applied to the crystalline state is described in the book *NMR Crystallography* (Harris et al (2009)) which surveys many examples of this; Chapter 27 by Middleton (2009) is devoted to structural biology applications. An overview of solid-state NMR in drug design and discovery for membrane-embedded targets is given by Watts (2005).

This chapter focuses on solution state NMR. A description of NMR crystallography was already given in Chapter 7.

The IUCr journal *Acta Cryst F Structural Biology Communications* at its launch worked closely with leading NMR spectroscopists to help define validation criteria (Einspahr and Guss (2008)). IUCr Journals include standard experimental tables for biological NMR in their notes for authors. A formatting for the tables can be found in their Word template at https://journals.iucr.org/services/wordstyle.html. Helpful notes and/or examples are given in the second column of Table 13.1. Examples of NMR protein structures published in the journal are by Serrano et al (2010) and Jaudzems et al (2010). Each of these two studies also included comparisons with the respective X-ray protein crystal structures, which nicely highlighted the complementarities of the two methods to study protein structure and dynamics.

At the PDB there are also helpful guidelines for depositions of NMR structures and many examples. A simple keyword search on one of the historically important protein solution structures determined by NMR 'BPTI NMR' (see Wüthrich and Wagner (1975) and Wüthrich (2002), and references therein) yields 1JV8 (https://www.rcsb.org/structure/1JV8 and Figure 13.1).

Precision and Accuracy in Biological Crystallography, Diffraction, Scattering, Microscopies, and Spectroscopies. John R. Helliwell, Oxford University Press. © John R. Helliwell (2025). DOI: 10.1093/9780198952848.003.0013

Table 13.1 NMR and refinement statistics

Distance constraints	**Notes/examples**
Total NOE	
Intra-residue	
Inter-residue	
Sequential ($\|i-j\| = 1$)	
Non-sequential ($\|i-j\| >1$)	Include as appropriate for complexes
Medium range ($\|i-j\| < 4$)	
Long range ($\|i-j\| = > 5$)	
Intermolecular	Include protein–nucleic acid intermolecular as appropriate
Hydrogen bond constraints	
Total dihedral angle restraints	
Protein	
φ	
ψ	
Nucleic acid	Include as appropriate
Base pair	
Sugar pucker	
Backbone	
Based on A-form geometry	
Residual dipolar coupling (RDC) constraints	Include if used
Total	
Q_{free}	
Structure statistics	
Violations (mean and s.d.)	
Distance constraints (Å)	
Dihedral angle constraints (°)	
Max. dihedral angle violation (°)	
Max. distance constraint violation (Å)	
Deviations from idealized geometry	
Bond lengths (Å)	
Bond angles (°)	
Impropers (°)	
Average pairwise r.m.s. deviation (Å)	Please indicate the number of structures used in r.m.s. deviation calculations
Protein	
Heavy atoms	
Backbone	
Nucleic acid	Include as appropriate
All nucleic acid heavy atoms	
Nucleic acid binding site	
All nucleotides	e.g. special parts, apical, tetraloop, symmetric bulge

Table 13.1 *continued*

Distance constraints	**Notes/examples**
Complex	
All complex heavy atoms (C, N, O, P)	
Protein and nucleic acid heavy atoms	
Ramachandran plot	Cite the source of restraints/definitions
Favoured regions (%)	
Outliers (%)	
Unmodelled/incomplete residues (%)	

The above Table is quoted from the *Acta Cryst F* Notes for Authors with the permission of IUCr Journals.

Figure 13.1 Ribbon diagram of the NMR structure of BPTI Mutant G37A (1JV8) (Battiste et al (2002)) coloured by sequence number. Image saved from the PDB 1JV8 entry https://www.rcsb.org/structure/1JV8. The ensemble in this case comprises 23 structures as fits to the NMR experimental data.

Figure 13.1 makes it obvious that the PDB deposition is an ensemble of structures as fits to the NMR experimental data. Wüthrich's (2002) Nobel Lecture (and references therein) neatly describes the model fit procedure. Wüthrich (2002) also makes clear that the NMR structure determination is 'precise'. Regions of the protein where there is a '*tight fit of the bundle indicates that the structure is defined with high precision, whereas the two chain ends are disordered*'. Furthermore, Wüthrich (2002) describes estimating the precision as follows: '*The average of the pairwise root-mean-square distances (RMSD) calculated for a bundle of conformers is then taken as a measure for the precision of the structure determination*'. Obviously, these definitions are entirely in accord with the preferences for describing precision and accuracy described in this book.

The PDB Validation Report for an NMR protein structure is summarized in the 'slider diagram'. For example, Figure 13.2 shows the slider diagram for 1JV8.

Obviously, an atomic coordinates' error estimate of precision added to the PDB Validation Report, for both NMR and crystallography, would be an improvement in my view.

1 Overall quality at a glance ⓘ

The following experimental techniques were used to determine the structure:
SOLUTION NMR

The overall completeness of chemical shifts assignment was not calculated.

Percentile scores (ranging between 0 and 100) for global validation metrics of the entry are shown in the following graphic. The table shows the number of entries on which the scores are based.

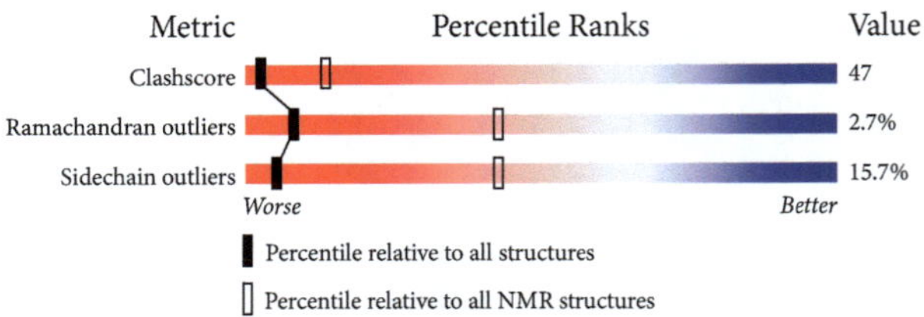

Figure 13.2 PDB slider diagram for 1JV8 from the PDB Validation Report, showing an 'Overall quality at a glance' of the structure, in this case the protein shown in Figure 13.1. There are three slider bars; to the left means a worse than average fit, and to the right a better than average fit. Markers on each of the three slider bars are either filled in, representing this structure relative to all deposited structures, or empty, representing this structure relative to all deposited NMR structures.

Key Learning Points

- NMR is a mature methodology in structural biology.
- *Acta Cryst F Structural Biology Communications* has published a variety of NMR studies.
- Some further articles in *Acta Cryst F* have cross-validated NMR and X-ray crystallography, one to the other and vice versa. These articles illustrate the complementarities of the two structural probes.
- The PDB Validation Report and IUCr's notes for authors for publishing NMR structures are described in this chapter.

References

Battiste, J. L., Li, R., and Woodward, C. (2002). *A highly destabilizing mutation, G37A, of the bovine pancreatic trypsin inhibitor retains the average native conformation but greatly increases local flexibility.* Biochemistry, 41 (7), 2237–2245. doi: 10.1021/bi011693e

Einspahr, H. and Guss, M. (2008). *Validation of macromolecular structures: updating standards for publication of NMR structures in an IUCr journal.* Acta Cryst., F64, 63. https://journals.iucr.org/services/nmr/nmrmeetingsummary.html

Harris, R. K., Wasylishen, R. E., and Duer, M. J. (Eds). (2009). *NMR Crystallography.* John Wiley, Chichester, UK.

Jaudzems, K., Geralt, M., Serrano, P., Mohanty, B., Horst, R., Pedrini, B., Elsliger, M.-A., Wilson, I. A., and Wüthrich, K. (2010). *NMR structure of the protein NP_247299.1: Comparison with the crystal structure.* Acta Cryst., F66, 1367–1380.

Middleton, D. (2009). *Structural Biology.* Chapter 27 in Harris, R. K., Wasylishen, R. E., and Duer, M. J. (Eds), *NMR Crystallography.* John Wiley, Chichester, UK.

Serrano, P., Pedrini, B., Geralt, M., Jaudzems, K., Mohanty, B., Horst, R., Herrmann, T., Elsliger, M.-A., Wilson, I. A., and Wüthrich, K. (2010). *Comparison of NMR and crystal structures highlights conformational isomerism in protein active sites.* Acta Cryst., F66, 1393–1405.

Watts, A. (2005). *Solid-state NMR in drug design and discovery for membrane-embedded targets.* Nat Rev Drug Discov, 4, 555–568. https://doi.org/10.1038/nrd1773

Wüthrich, K. (2002). *NMR studies of structure and function of biological macromolecules.* Nobel Lecture, https://www.nobelprize.org/uploads/2018/06/wutrich-lecture.pdf.

Wüthrich, K. and Wagner, G. (1975). *NMR investigations of the dynamics of the aromatic amino acid residues in the basic pancreatic trypsin inhibitor.* FEBS Lett., 50, 265–268.

14
EPR for metalloproteins

Electron Paramagnetic Resonance (EPR) is a magnetic resonance technique which detects the resonance transitions between energy states of unpaired electrons in an applied magnetic field. Such an unpaired electron has spin, which gives it a magnetic moment. By contrast NMR detects resonance transitions between unpaired nuclear spins. Because of electron–nuclear mass differences, the magnetic moment of an electron is substantially larger than the corresponding quantity for any nucleus. This means that a much higher electromagnetic frequency is needed to bring about a spin resonance with an electron than with a nucleus, at identical magnetic field strengths. For applications to metalloenzymes and metals in medicine see Hanson and Berliner (2009).

EPR is a non-invasive (i.e. non-damaging) method as compared with X-ray crystallography or cryoEM. Whilst X-ray absorption spectroscopy is lower dose compared with X-ray crystallography, as highlighted earlier in the XAS chapter, it is still invasive, i.e. involves an absorbed X-ray dose compared with EPR (or NMR). EPR has found application with respect to a range of metalloproteins. It has found particular application to studies of PS II as a complement to X-ray crystallography (see Haddy (2007) for a review).

An influential paper in the PSII field is that of Cox et al (2014) who asserted that, using EPR, they 'confirm a hypothesis that all four Mn ions are octahedrally coordinated and in the 4+ oxidation state'. This has been challenged by Wang (2024) 'when generating structures from crystallographic data, one should ignore whatever inferences may have been drawn about them using non-crystallographic techniques if possible because they could severely compromise the independence of the conclusions ... The structural changes that occur as the [oxygen-evolving complex] OEC progresses through its duty cycle are small. The resolution of the crystal structures available for it are respectable for a macromolecule, but not all that one might like for a system this demanding. Lastly, its spectroscopic properties have been under investigation for many decades and the interpretation of these data has depended heavily on whatever one believes the underlying structure might be.'

Precision and Accuracy in Biological Crystallography, Diffraction, Scattering, Microscopies, and Spectroscopies.
John R. Helliwell, Oxford University Press. DOI: 10.1093/9780198952848.003.0014

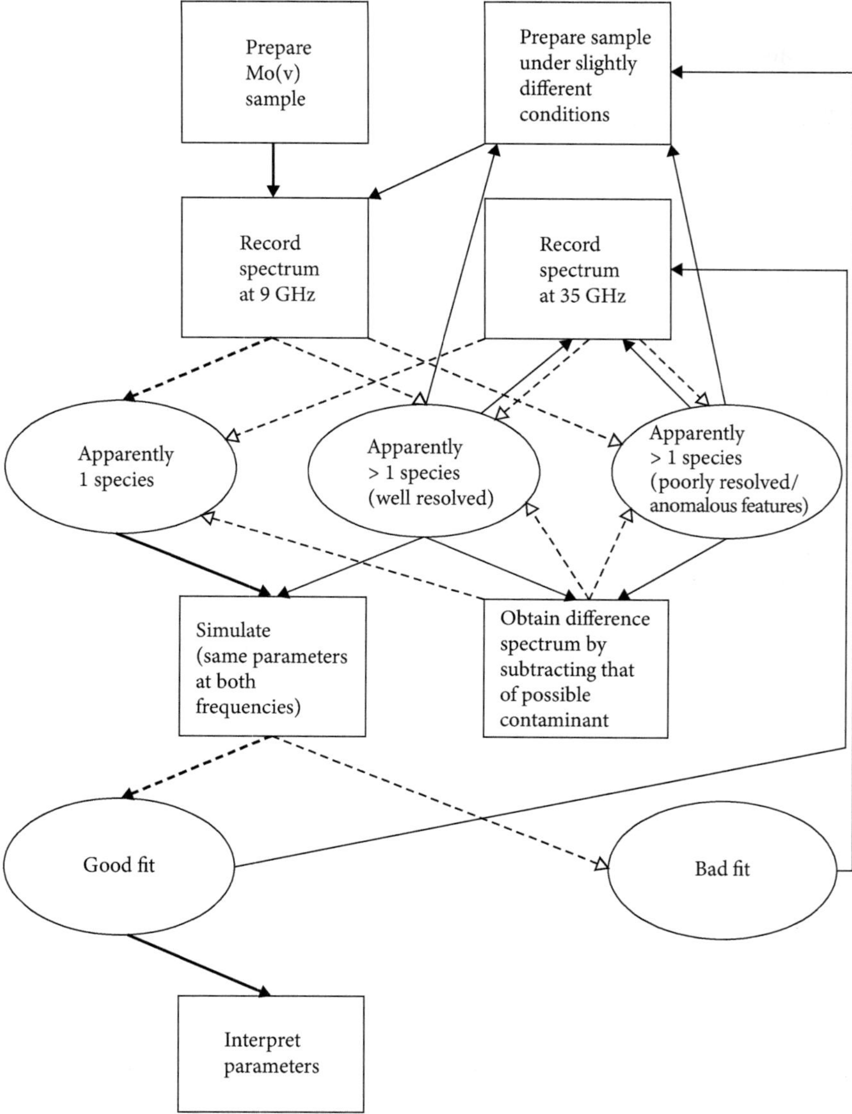

Figure 14.1 Scheme for deconvoluting overlapping EPR spectra, by using difference techniques, two microwave frequencies, and computer simulations. From the top, it shows a workflow of steps and decisions starting with sample preparation and proceeding through experimental measurements to an assessment of fit, culminating finally in the best parameters interpretation.

From R. C. Bray (1988). *The inorganic biochemistry of molybdoenzymes.* Quarterly Reviews of Biophysics, 21(3), 299–329. doi:10.1017/S0033583500004479 with permission of Cambridge University Press, 2009 © Cambridge University Press.

The above text quotation is with the permission of Dr Wang and the IUCr Journals.

For other studies, and whilst it is a review from 1988, Bray (1988) gives a nice description of the use of EPR to characterize molybdoenzymes (Figure 14.1 shows the practical procedure followed).

Bray (1988) also compared the efficacy of EPR and EXAFS in studying these enzymes, with his preference for EPR: 'Extracting structural information from EPR and EXAFS spectra are complex processes, in which comparison with appropriate model compounds plays a part.' Furthermore, EPR is sensitive to the presence of mixtures, which defies analysis by EXAFS (and, one might add, can make it challenging for X-ray protein crystallography): 'When applying e.p.r. to molybdoenzymes it is the rule, rather than the exception, for more than one signal-giving species of Mo(V) to be present and for these to have overlapping spectra.'

As further examples, Pushie and George (2011) give a detailed description and evaluation of the structural studies of molybdenum and tungsten enzymes made by X-ray crystallography, EXAFS, and EPR. EPR can also yield distances between unpaired electron centres (see e.g. Schiemann and Prisner (2007)).

Key Learning Points

- EPR is a non-invasive technique applicable to biological systems where there are unpaired electrons.
- EPR has been used to study the oxygen-evolving complex in PSII. Interpretation of those data has been remarked upon.
- With reference to molybdoenzymes, EPR and EXAFS results have been contrasted.
- Extensive reviews and books on the topic are available.

References

Bray, R. C. (1988). *The inorganic biochemistry of molybdoenzymes.* Quarterly Reviews of Biophysics, 21(3), 299–329. doi:10.1017/S0033583500004479

Cox, N. et al (2014). *Electronic structure of the oxygen-evolving complex in photosystem II prior to O-O bond formation.* Science, 345, 804–808. doi:10.1126/science.1254910

Haddy, A. (2007). *EPR spectroscopy of the manganese cluster of photosystem II.* Photosynth Res, 92, 357–368.

Hanson, G. and Berliner, L. (Eds) (2009). *High Resolution EPR: Applications to Metalloenzymes and Metals in Medicine.* Springer-Verlag, New York.

Pushie, M.J. and George, G.N. (2011). *Spectroscopic studies of molybdenum and tungsten enzymes.* Coordination Chemistry Reviews, 255, 1055–1084.

Schiemann O. and Prisner T.F. (2007). *Long-range distance determinations in biomacromolecules by EPR spectroscopy.* Quarterly Reviews of Biophysics, 40(1), 1–53. doi:10.1017/S003358350700460X

Wang, J. (2024). *Photosystem II: Light-dependent oscillation of ligand composition at its active site.* Acta Cryst., D80. 850–861.

15
Combining methods for accuracy

As remarked several times earlier in this book, individual methods have systematic errors. Accuracy can potentially be achieved by utilizing at least two distinct methods for structure determination, each with different systematic errors. As an example, X-ray crystallography and NMR structures of the same protein were compared in detail in Chapter 13. Combining methods studying solution and crystalline states can also show up major differences in conformation of a biological macromolecule, such as reported by Wüthrich (1986) for glucagon using NMR, as well as calmodulin and troponin C1 using SAXS (Heidorn and Trewhella (1988)).

There are situations where the 3D structure of a biological macromolecule is not strictly essential to understand a biological process and, likewise, neither is the accuracy or precision of the atomic coordinates strictly necessary. This would be where an enzyme function assay shows that an additive inhibits a reaction by the enzyme, i.e. slows it down or stops it altogether. It is nevertheless always useful to know where that inhibitor has bound to know if it acts directly at the active site or at an allosteric binding site. In these cases, if the 3D structure isn't determined at body temperature, cryo-artefacts such as those identified by Halle (2004), especially low occupancy binding, also don't really matter because the enzyme assay in solution at a physiologically relevant temperature confirms or not an additive's inhibitory effect.

A careful analysis by Wang (2024) provides a cautionary example of the pitfalls of combining before even comparing structural results from two methods: in this case by constraining and overfitting manganese ions in the oxygen-evolving centre of PSII.

In Case study 15.1 two methods are combined, X-ray and neutron macromolecular crystallography. A weakness of this combination, of course, is that both use the macromolecule in a crystalline state, and usually the same crystal form, i.e. space group, and hence that while the data are complementary, any structural artefact arising from the crystal form will be observed in both studies. Case study 15.2 is more wide ranging in that the NMR and the SAXS are both measurements on the dynamic solution state of malate synthase G.

Precision and Accuracy in Biological Crystallography, Diffraction, Scattering, Microscopies, and Spectroscopies.
John R. Helliwell, Oxford University Press. © John R. Helliwell (2025). DOI: 10.1093/9780198952848.003.0015

That case study also describes independent determinations of the macromolecular structure of malate synthase G by X-ray crystallography and cryoEM, and how the results compare.

15.1 Combining X-ray crystallography then neutron crystallography

First to be mentioned, and most notably, macromolecular X-ray crystallography, even at high diffraction resolution, cannot precisely locate the hydrogen atoms. This is because the electron density of the covalent bond is away from the proton of the hydrogen atom. This may be regarded as a somewhat special case where very high resolution is available. Nevertheless, neutrons do reveal the nuclear scattering of the proton and thereby its correct position (Gruene et al (2014)).

At medium resolution (e.g. 2Å) very few if any hydrogen atoms are located by X-ray crystallography. What method can it be combined with to circumvent this systematic error? Most notably the method of choice is macromolecular neutron crystallography as it is sensitive to deuterium and is non-radiation damaging. This method needs as a prerequisite an X-ray crystal structure and, ideally, a fully deuterated macromolecule (and if relevant a fully deuterated ligand). Single-particle cryoEM can locate hydrogens, being sensitive to the electrostatic potential, which is superior to the X-ray atomic scattering factor for hydrogen but can cause greater radiation damage. Hence the absence of a hydrogen on an ionizable amino acid side chain cannot necessarily be attributed to be its natural chemical state in a cryoEM structure. Note that functional moieties of ionizable side chains, particularly acidic ones, are less visible in cryoEM maps: as Bartesaghi et al (2014) remark, 'Inspection of the map reveals that although densities for residues with positively charged and neutral side chains are well resolved, systematically weaker densities are observed for residues with negatively charged side chains'.

Finally, as a complementary method, what about prediction? The position of hydrogens on the non-ionizable part of a macromolecule can be confidently predicted because of their various, known, geometrical stereochemistries. The prediction of the protonation states of Asp, Glu, His, Lys, and Arg in a protein can be attempted as there are various software packages for trying to do this. Evaluation of these should be kept under review as the prediction software may improve; one such evaluation is that of Fisher et al (2009).

15.2 The case of malate synthase G studied by MX, NMR, SAXS, and cryoEM

This enzyme comprises 723 amino acids, i.e. ~80 kDa molecular weight. It is, then, rather large for 'routine NMR' and rather small for 'routine cryoEM'. The studies on the enzyme by NMR attracted the focus of SAXS specialists, who assessed how SAXS could assist NMR in reaching an improved structure determination in solution. Altogether these individual structures from each of these several methods are compared below, so as to evaluate how well they agree. This, then, will indicate how each method sits with respect to the overall accuracy.

Firstly Figure 15.1 (Figure 2 of Ho et al (2023)) compares cryoEM, X-ray (PDB code 1D8C), and NMR malate synthase G structures. It also tabulates the rms deviations between the structures. These can be compared with the Cruickshank Diffraction Precision Index (DPI) from the Online DPI web-server (Kumar et al (2015)) for 1D8C which is, using R_{work} (17.5%), 0.16 (Å), i.e. certainly smaller than the rms deviation compared values between these various methods. Also, Ho et al (2023) emphasize that the crystallographic and cryoEM structures both show two helices shifted considerably (12 Å) relative to the NMR solution structure (see Figure 15.1C).

Ho et al (2023) concluded as follows:

> *In this work, we have demonstrated that the structure of 723-residue ecMCG can be well determined by cryo-EM technology. The overall structures and details of the residue sidechains of the protein structures resolved by both cryo-EM and X-ray crystallography are significantly identical, implying that the cryo-EM structure of ecMSG we present herein is reliable. Our high-quality ecMSG cryo-EM structure validates the feasibility of routinely deploying cryo-EM for structural determinations of sub-100 kDa proteins.*

This last remark in the quote just above is an interesting emphasis in that Ho et al (2023) treat the X-ray crystal structure as the benchmark for their cryoEM structure. As their second main conclusion they compare the structural dynamics:

> *Dynamic ecMSG variabilities derived from cryo-EM analysis, crystallographic ADPs and NMR order parameters are highly consistent, evidencing the reliability of structural heterogeneities illustrated by cryo-EM.*

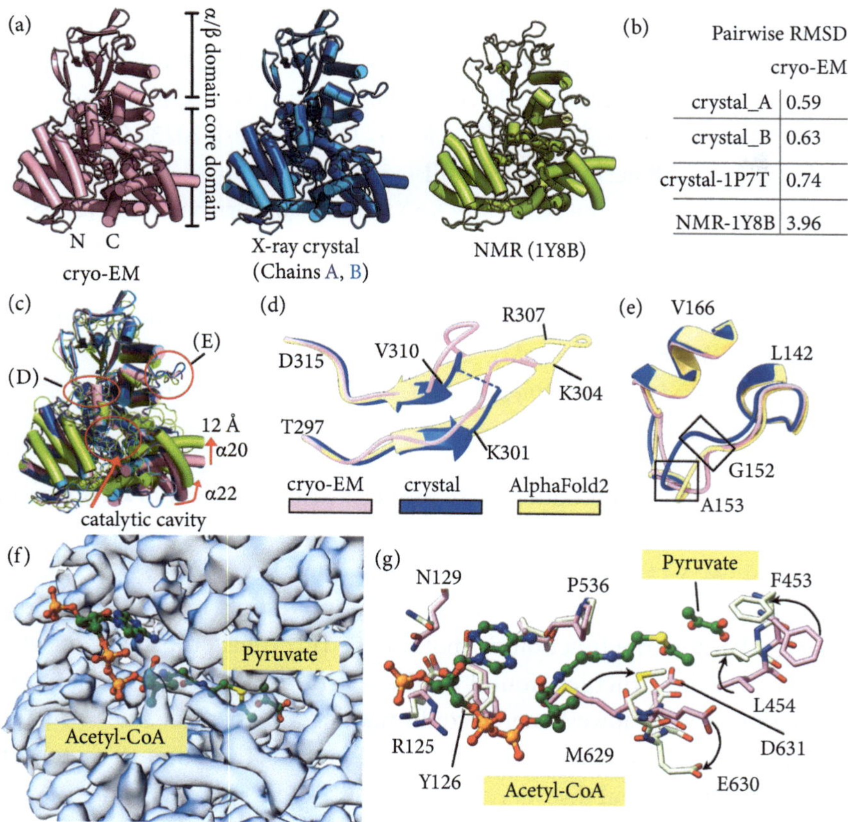

	cryo-EM
crystal_A	0.59
crystal_B	0.63
crystal-1P7T	0.74
NMR-1Y8B	3.96

Figure 15.1 Comparison of apo-form ecMSG structures by different methods. (A) Alignments and superimpositions of apo-form ecMSG structures determined by cryoEM (pink), X-ray crystallography (cyan and blue), and solution NMR spectroscopy (green). Globally, the determined protein folding is similar amongst methodologies, but the NMR structure ensemble is more dynamic. (B) Calculated rmsds (Å) of global structural alignments over 712 residues (C_α) paired to the cryoEM ecMSG structure. (C) The aligned cryoEM, crystal, and NMR structures of apo-ecMSG illustrate that the α20 and α22 helices of the NMR structure are shifted ~ 12 Å toward the α/β-domain, whereas the cryoEM and crystal structures are identical, clearly a very significant finding. Circles indicate variations analysed in panels D, E, and F. (D) Localized structure of ecMSG residues 297–315 are distinctive among the three different methodologies, including Alphafold2 (yellow) (Jumper et al (2021)), suggesting a dynamic feature. (E) A loop (residues 142–160) between helices α6 and α7 is also dynamic. Residues G152 and A153 are labelled to highlight divergence among the three structures. (F) Aligned acetyl-CoA and pyruvate from the crystal structure (1P7T) reveals an unoccupied space in the

(this caption continues at the foot of page 90)

The two text extracts above are reproduced from Meng-Ru Ho, Yi-Ming Wu, Yen-Chen Lu, Tzu-Ping Ko and Kuen-Phon Wu (2023), *Cryo-EM reveals the structure and dynamics of a 723-residue malate synthase G*, Journal of Structural Biology, 215, 107958, https://doi.org/10.1016/j.jsb.2023.107958 Copyright (2024), with permission from Elsevier.

Moving on now to the NMR and SAXS studies by Grishaev et al (2008):

> *For the structure calculation beyond (using) the NMR-only ... including low-angle SAXS data up to a q_{max} of 0.220 Å^{-1} produced a considerable improvement of the structural quality, as manifested by a 0.85 Å decrease of the backbone rmsd to the full-length (residues 3–722) X-ray structure of MSG (PDB code 1D8C).*

This text quotation reproduced with the permission of Professor Grishaev and Springer Nature.

Again, an X-ray crystal structure (1D8C here) is used as the reference benchmark. This in itself is an especially interesting insight into current practice because the functioning molecular structure is probably somewhere in between the structures found by the several methods used. The crystal structure gives the lowest energy conformation and does not show structural effects of dynamics (albeit there are indications from the atomic B-factors and modellable positional disorder). Then cryoEM is far more dependent on the skill of the user to get the best fit of the protein model onto the electrostatic potential map. The use of SAXS and NMR could be very helpful as a combination, depending on how that is implemented; the NMR relies on sufficient interatomic and distance restraints to get the tertiary fold correct (Figure 15.2).

Grishaev et al (2008) concluded as follows:

> *In this study we observed a substantial improvement in structural accuracy for an 82 kDa protein when orientational NMR restraints obtained in a weakly aligning liquid crystalline medium are supplemented by SAXS data in a direct*

Figure 15.1 (continued) catalytic pocket of the apo-form cryoEM map. (G) Sidechain comparisons of the apo- (cryoEM, pink) and complex-form ecMSG (1P7T, light green) structures illustrate that residues F453 and L454 flip and stably clamp the pyruvate (arrows). Residues M629 and E630 are oriented outward to provide space for hosting the carbon chain of acetyl-CoA (arrow).
Reproduced from Meng-Ru Ho, Yi-Ming Wu, Yen-Chen Lu, Tzu-Ping Ko and Kuen-Phon Wu (2023), *Cryo-EM reveals the structure and dynamics of a 723-residue malate synthase G*, Journal of Structural Biology, 215, 107958, https://doi.org/10.1016/j.jsb.2023.107958. Copyright (2024), with permission from Professor K.-P. Wu and Elsevier.

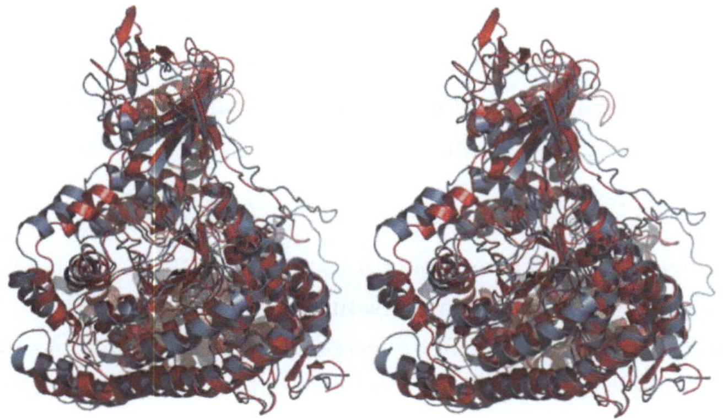

Figure 15.2 Stereo representation of the superimposition of the structure of malate synthase G (MSG) obtained by the joint fit of SAXS and NMR data (red) and the NMR-only model (blue). The figure is generated with PyMOL (DeLano 2002) from Grishaev et al (2008).
Reproduced with the permission of the authors and Springer Nature.

refinement procedure ... The SAXS data improved the structural accuracy by influencing the relative positioning of the individual domains within the overall model. The bulk of this improvement does not result from either the inner-most part of the scattering curve, which is sensitive to the overall particle dimension, or from the higher-angle data, which are influenced by variation of the protein's internal density. Instead, the intermediate angle data, which reflect the low-resolution particle shape, are the driving force behind the rearrangement of the MSG domains.

Grishaev et al (2008) are also careful about the differences between the methods involved:

A joint analysis of NMR and scattering data indicates that the magnitude of the differences between the MSG domain orientations in solution and in the crystal is only slightly higher than the experimental uncertainty of the relative domain orientations. No statistically significant translational displacements of the peripheral domains with respect to those in the X-ray structure of the glyoxylate bound MSG were detected from our scattering data.

Their main thrust is the potential of combining the two methods of NMR and SAXS on large macromolecules like MSG in solution:

In this study, immediately analysable SAXS data were acquired in less than 1 hour, using a small fraction of the NMR sample at a moderate concentration.

Data analysis tools that we employed are user-friendly and straightforward even for a novice. We thus expect that the application of the joint NMR/SAXS structure refinement methodology will become routine for other larger macromolecular systems, where acquisition of NMR structural restraints is particularly challenging.

The above figure and three text quotations are reproduced with the permission of Professor Grishaev and Springer Nature.

The crystal structure of 1D8C referred to above was of MSG in complex with Mg^{2+} and glyoxylate at 2.0 Å resolution and undertaken by Howard et al (2000).

15.3 Simultaneous measurements made from the same sample

An example of this is the small-angle scattering instrument at the Institut Laue Langevin D22 where the same sample is studied simultaneously by SAXS and SANS as well as a UV/Vis spectrophotometer (see Figure 15.3); further details are in Metwalli et al (2020, 2021). An example of applying this simultaneous combination of probes is for studying lipid nanoparticles (LNPs) used in formulations for mRNA drug delivery (Unruh et al (2024)). Using this approach both SAXS and SANS data from the same sample volume can be fitted simultaneously using a common structural model. Also, *in situ*/real-time investigations are possible, where the simultaneous SAXS and SANS data are used in gaining time-resolved complementary nanoscale structural information.

Key Learning Points

- Comparing methods in structural biology is very informative. Several examples are provided in detail in case studies in this chapter and in Chapter 13.
- Rms differences between structures can be compared with the Cruickshank (1999) estimate for an X-ray crystal structure.
- Simultaneous measurements on the same sample can be feasible, such as SAXS with SANS as well as UV/Vis spectroscopy in the case study described, but it is an approach that is in its infancy.

Figure 15.3 Simultaneous SAXS and SANS on the same sample. Top view: Left to right is the X-ray beam from a sealed tube X-ray source and forward to back is the direction of the neutron beam. Middle view: Same view as top view but zoomed in to see the sample stage. Bottom view is of the UV/Vis spectrophotometer.

Figures kindly provided by Prof Dr Tobias Unruh.

References

Bartesaghi, A., Matthies, D., Banerjee, S., Merk, A., and Subramaniam, S. (2014). *Structure of β-galactosidase at 3.2-Å resolution obtained by cryo-electron microscopy.* Proceedings of the National Academy of Sciences, 111(32), 11709–11714.

Cruickshank, D. W. J. (1999). *Remarks about protein structure precision.* Acta Cryst., D55, 583–601.

DeLano, W. L. (2002). *The PyMOL molecular graphics system.* DeLano Scientific, Palo Alto, CA, USA.

Fisher, S. J., Wilkinson, J., Henchman, R. H., and Helliwell, J. R. (2009). *An evaluation review of the prediction of protonation states in proteins versus crystallographic experiment.* Crystallography Reviews, 15, 231–259.

Grishaev, A., Tugarinov, V., Kay, L. E., Trewhella, J., and Bax, A. (2008). *Refined solution structure of the 82-kDa enzyme malate synthase G from joint NMR and synchrotron SAXS restraints.* Journal of Biomolecular NMR, 40(2), 95–106. doi: 10.1007/s10858-007-9211-5. PMID: 18008171

Gruene, T., Hahn, H. W., Luebben, A. V., Meilleur, F., and Sheldrick, G. M. (2014). *Refinement of macromolecular structures against neutron data with SHELXL2013.* J. Appl. Cryst., 47, 462–466.

Halle, B. (2004). *Biomolecular cryocrystallography: Structural changes during flash-cooling.* Proc Natl Acad Sci U S A., 101 (14), 4793–4798. doi: 10.1073/pnas.0308315101. Epub 2004 Mar 29. PMID: 15051877; PMCID: PMC387327

Heidorn, D. B. and Trewhella, J. (1988). *Comparison of the crystal and solution structures of calmodulin and troponin C.* Biochemistry, 27 (3), 909–915.

Ho, M.-R., Wu, Y.-M., Lu, Y.-C., Ko, T.-P., and Wu, K.-P. (2023). *Cryo-EM reveals the structure and dynamics of a 723-residue malate synthase G.* Journal of Structural Biology, 215, 107958. https://doi.org/10.1016/j.jsb.2023.107958

Howard, B. R., Endrizzi, J. A., and Remington, S.J. (2000). *Malate synthase G complexed with magnesium and glyoxylate.* Biochemistry, 39, 3156–3168.

Jumper, J., Evans, R., Pritzel, A., Green, T., Figurnov, M., Ronneberger, O., Tunyasuvunakool, K., Bates, R., Žídek, A., Potapenko, A., Bridgland, A., Meyer, C., Kohl, S. A. A., Ballard, A. J., Cowie, A., Romera-Paredes, B., Nikolov, S., Jain, R., Adler, J., Back, T., Petersen, S., Reiman, D., Clancy, E., Zielinski, M., Steinegger, M., Pacholska, M., Berghammer, T., Bodenstein, S., Silver, D., Vinyals, O., Senior, A. W., Kavukcuoglu, K., Kohli, P., and Hassabis, D. (2021). *Highly accurate protein structure prediction with AlphaFold.* Nature, 596, 583–589.

Metwalli, E., Götz, K., Lages, S., Bär, C., Zech, T., Noll, D. M., Schuldes, I., Schindler, T., Prihoda, A., Lang, H., Grasser, J., Jacques, M., Didier, L., Cyril, A., Martel, A., Porcar, L., and Unruh, T. (2020). *A novel experimental approach for nanostructure analysis: Simultaneous small-angle X-ray and neutron scattering.* J. Appl. Cryst., 53, 722–733.

Metwalli, E., Götz, K., Zech, T., Bär, C., Schuldes, I., Martel, A., Porcar, L., and Unruh, T. (2021). *Simultaneous SAXS/SANS method at D22 of ILL: Instrument upgrade.* Appl. Sci., 11, 5925. https://doi.org/10.3390/app11135925

Unruh, T., Götz, K., Vogel, C., Fröhlich, E., Scheurer, A., Porcar, L., and Steiniger, F. (2024). *Mesoscopic structure of lipid nanoparticle formulations for mRNA drug delivery: Comirnaty and drug-free dispersions.* ACS Nano, 18 (13), 9746–9764.

Wang, J. (2024). *Photosystem II: light-dependent oscillation of ligand composition at its active site.* Acta Cryst., D80, 850–861.

Wüthrich, K. (1986). *Glucagon conformation in different environments: Implications for molecular recognition.* In: Van Binst, G. (ed.), *Design and Synthesis of Organic Molecules Based on Molecular Recognition.* Springer, Berlin, Heidelberg. https://doi.org/10.1007/978-3-642-70926-5_6.

16

Combining methods to span different length scales

Overall, there is scepticism of what we 'atomic level structuralists' do for biology:

Dame Ottoline Leyser, Professor of Plant Development at the University of Cambridge, was quoted (Turney (2019)) stating that:

> *The defining feature of biology during the past few decades has been figuring out details of the parts. But biological systems don't think they have parts.*

Scepticism of the role of reductionism in understanding biology is also a theme in Ernst Mayr's book *What makes biology unique?* (Mayr (2007)). Mayr even argues against the relevance of the discovery of the DNA double helix to understanding biology.

As a counterpoint, a founding father of quantum mechanics, the physicist Erwin Schrödinger posed the question 'What is Life?' in his influential book *What is Life?: The Physical Aspect of the Living Cell*, in effect applying the physical sciences to this central question of biology.

But then we have Sir Paul Nurse, awarded the Nobel Prize in Physiology or Medicine in 2001 for his work on the cell cycle of fission yeast (interviewed by Ireland (2014)):

> *Rather than get bogged down in detailed molecular descriptions of everything, I'm asking bigger questions like 'how does a cell know how big it is?', which has always fascinated me.*

So, where do these strong views of the holistic biologists stand today against our reductionist research in the molecular sciences, within which crystallography is a key player, as are the microscopies and spectroscopies. More to the point, how can the crystallographer respond constructively, maybe only partly, to the concerns of the holistic biologists? Let's first précis arguments given elsewhere in this book about accuracy, including functionally relevant (to the living cell) macromolecular structures.

Precision and Accuracy in Biological Crystallography, Diffraction, Scattering, Microscopies, and Spectroscopies. John R. Helliwell, Oxford University Press. © John R. Helliwell (2025). DOI: 10.1093/9780198952848.003.0016

The issue of relevance starts when considering the crystalline state or the solution state of a protein, but imagined placed inside the biological cell. NMR provides atomically detailed results in solution and of course protein crystallography provides atomic details for a protein in the solid state. The protein crystal is, though, a curious hybrid of solid state, being an ordered lattice, yet with a large percentage of the crystal volume comprising solvent channels that run through the crystal.

Studies of the structure and function of an enzyme in the crystalline state were to my mind greatly facilitated by the invention of the flow cell (Wyckoff et al (1967)). This is a powerful approach to the issue of the relevance of our protein crystal results. It answers with a resounding yes that these results are relevant to enzyme function and overcomes the objections of the NMR solution state spectroscopists to the crystallographer's results in the protein crystal. Fine examples of direct 'catalysis in the crystal' were undertaken on glycogen phophorylase b (Hajdu et al (1987)). The direct monitoring of changes of UV/Vis signals directly from the crystal became an important supplement to the fast X-ray crystallography studies by invention of the microspectrophotometer (Hadfield and Hajdu (1993)). This was widely applied including deciding when to freeze trap an intermediate state (Bourgeois and Weik (2009).

Weaknesses in the armoury of crystallography remain, such as crystallization conditions, to a greater or lesser degree, taking one's results away from biological functioning conditions. As Yibin Lin's article (Lin (2018)) states:

> *scientists frequently select the protein that is suitable for crystallization but far from the physiological condition*

This criticism of Yibin Lin (2018) is perhaps more relevant than the one about having a lattice for a protein crystal. Lattice contacts in macromolecular crystals are rather few, whereas other possible effects of non-physiological crystallization conditions cannot be similarly dismissed as they affect the surface chemistry of the atoms in the crystallized protein. By this I mean that the folds of any particular protein often superimpose closely regardless of crystallization conditions or unit cell. More specific effects of pH on protonation states or some ligands are the examples of surface chemistry to which I think Lin (2018) refers.

A similar concern is the question about the strict relevance to biology of crystallography results now predominantly based on X-ray diffraction data measured at cryo-temperatures (Halle (2004)). This has been compounded by observations of specific X-ray damage to the crystallized protein. Conducting crystallography at physiologically relevant temperatures has then become an objective. Jacobs et al (2024), for medical applications, define this as body temperature (37 °C) protein crystallography. Of course, with X-rays radiation damage will increase at body temperature; this can be avoided by use of serial femtosecond crystallography or neutron crystallography.

Neutron macromolecular crystallography (nMX) has automatically yielded room temperature structures. Given also that projects succeed in getting beamtime only where all other methods have failed (X-ray, electron or NMR based), it is clear that in structural biology there is a strategic importance of the nMX method. nMX has seen a sustained growth in both instruments and development of software and methods; see Chapter 7 and examples in the comprehensive recent text (Moody (2020)).

The new femtosecond range X-ray lasers yield X-ray diffraction data at room temperature before radiation damage can kick in. Synchrotron facilities are now also adopting the X-ray laser methods for delivery of streams of micron-sized samples and thereby are also yielding results at room temperatures, albeit not completely free of radiation damage like the X-ray lasers. The use of streams of micron-sized crystals may be susceptible to variability within samples of the biological molecules being studied.

So, the physical methods of crystallography, microscopy, and spectroscopy continue to strive for, and do clearly deliver, functionally relevant structural results. While many biologists have embraced molecular structure shall we ever convince holistic biologists? Our studies rarely take us directly to the whole organism. In Chapter 17, I present a case study where that all-important bridge to whole animals is achieved, the coloration of the shell of live lobsters.

Now let's come to the wider issue of what we do at the molecular level, at the nanoscale level, and up to the cellular level of scale? This is the work of integrative structural biology (Ward et al (2013), Sali et al (2015)). Table 16.1 shows the range of data, and the methods used, to facilitate integrative modelling.

A rapidly developing method is that of cryo-electron tomography recently reviewed by Nogales and Mahamid (2024). As they neatly describe:

> *cryo-EM and cryo-electron tomography, as a connecting bridge to visualize macromolecules* in situ, *holds great promise to create comprehensive structural depictions of macromolecules as they interact in complex mixtures or, ultimately, inside the cell itself*

Table 16.1 Types of structural data and the methods used to facilitate integrative modelling.

Structural Information	Method
Atomic structures of parts of the studied system	X-ray and neutron crystallography, NMR spectroscopy, 3DEM, comparative modelling, and molecular docking
3D maps and 2D images	Electron microscopy and tomography
Atomic and protein distances	NMR, FRET, and other fluorescence techniques, DEER, EPR, and other spectroscopic techniques; chemical crosslinks detected by mass spectrometry, and disulfide bonds detected by gel electrophoresis
Binding site mapping	NMR spectroscopy, mutagenesis, FRET
Size, shape, and pairwise atomic distance distributions	SAS
Shape and size	Atomic force microscopy, ion mobility mass spectrometry, fluorescence correlation spectroscopy, and fluorescence anisotropy
Component positions	Super-resolution optical microscopy, FRET imaging
Physical proximity	Copurification, native mass spectrometry, genetic methods, and gene/protein sequence covariance
Solvent accessibility	Footprinting methods, including H/D exchange assessed by mass spectrometry or NMR, and even functional consequences of point mutations
Proximity between different genome segments	Chromosome conformation capture and other data
Propensities for different interaction modes	Molecular mechanics force fields, potentials of mean force, statistical potentials, and sequence covariation

Example methods that are informative about a variety of structural aspects of biomolecular systems are listed. 3DEM, 3D electron microscopy; DEER, double electron–electron resonance; EPR, electron paramagnetic resonance; FRET, Förster resonance energy transfer; H/D, hydrogen/deuterium; NMR, nuclear magnetic resonance; SAS, small-angle scattering.

Reproduced from Sali, A. et al (2015). *Outcome of the First wwPDB Hybrid/Integrative Methods Task Force Workshop.* Structure, 23(7), 1156–1167. Copyright (2015), with permission from Dr Andrej Sali and Elsevier.

The whole-cell view is beautifully linked together by modelling and even art, e.g. as depicted in https://gaelmcgill.artstation.com/projects/Pm0JL1.

Key Learning Points

- Molecular level researchers are very aware of requiring functional relevance and have explored this frontier by every means possible.
- Integrative structural biology is well recognized by researchers and the PDB in its deposition procedures.
- Visualization is a powerful means of depicting what happens in the living cell.

References

Bourgeois, D. and Weik, M. (2009). *Kinetic protein crystallography: A tool to watch proteins in action.* Crystallography Reviews,15 (2), 87–118.

Hadfield, A. and Hajdu, J. (1993). *A fast and portable microspectrophotometer for protein crystallography.* J. Appl. Cryst., 26, 839–842.

Hajdu, J., Acharya, K. R., Stuart, D. I., McLaughlin, P. J., Barford, D., Oikonomakos, N. G., Klein, H. W., and Johnson, L. N. (1987). *Catalysis in the crystal: Synchrotron radiation studies with glycogen phosphorylase b.* EMBO J., 6, 539–546.

Halle, B. (2004) *Biomolecular cryocrystallography: Structural changes during flash-cooling.* Proc. Natl. Acad. Sci. U.S.A., 101, 4793–4798.

Ireland, T. (2014). *Interview of Sir Paul Nurse.* The Biologist, 61 (6), 32–35.

Jacobs, F. J. F., Helliwell, J. R., and Brink, A. (2024). *Body temperature protein X-ray crystallography at 37 °C: A rhenium protein complex seeking a physiological condition structure.* Chemical Communications. doi: 10.1039/D4CC04245J

Lin, Y. (2018). *What's happened over the last five years with high-throughput protein crystallization screening?* Expert Opinion on Drug Discovery, 13, 691–695.

Mayr, E. (2007). *What makes biology unique?* Cambridge University Press, Cambridge.

Moody, P. (2020). *Neutron crystallography in structural biology* (Methods in Enzymology series, Vol. 634). Academic Press, Cambridge, Mass., USA.

Nogales, E. and Mahamid, J. (2024). *Bridging structural and cell biology with cryo-electron microscopy.* Nature, 628, 47–56. https://doi.org/10.1038/s41586-024-07198-2

Sali, A., Berman, H. M., Schwede, T., Trewhella, J., Kleywegt, G., Burley, S. K., Markley, J., Nakamura, H., Adams, P., Bonvin, A. M., and Chiu, W., 2015. *Outcome of the first wwPDB hybrid/integrative methods task force workshop.* Structure, 23 (7), 1156–1167.

Schrödinger, E. (2012). *What is Life?: The Physical Aspect of the Living Cell.* Based on lectures delivered under the auspices of the Dublin Institute for Advanced Studies at Trinity College, Dublin, in February 1943. Reprinted by Cambridge University Press, Cambridge, UK.

Turney, J. (2019). *The puzzle of life.* Times Higher Educational Supplement, 21 February 2019, pages 44–45.

Ward, A. B., Sali, A., and Wilson, I. A. (2013). *Integrative Structural Biology.* Science, 339, 913–915. doi: 10.1126/science.1228565

Wyckoff, H. W., Doscher, M., Tsernoglou, D., Inagami, T., Johnson, L. N., Hardman, K. D., Allewell, N. M., Kelly, D. M., and Richards, F. M. (1967). *Design of a diffractometer and flow cell system for X-ray analysis of crystalline proteins with applications to the crystal chemistry of ribonuclease-S.* J. Mol. Biol., 27, 563–578.

17

Role of simulations of structural dynamics as a complement to experimental studies

17.1 Molecular dynamics

As computer hardware has grown in capabilities over the decades, computational scientists have been very active in simulations of molecular dynamics (MD) in general and protein dynamics in particular. Perhaps a little old now, but a very useful review I find is that of McCammon (1984), and for a broad overview of molecular modelling see Leach (2001). The UK's collaborative computational project for biomolecular simulation is an active collaboration on the whole topic and runs training workshops, conferences, and supports software development (see https://www.ccpbiosim.ac.uk/).

MD simulations have come along in leaps and bounds over the past couple of decades, and many different types of calculation and forcefield and advanced sampling methods are used. Also, computational power is much greater than it was. The software my laboratory and our collaborators have used is AMBER (https://ambermd.org/index.php) (see e.g. Bradbrook et al (1998) and Pratap et al (2001)). A significant leap forward in length of time of a simulation was that of Wall et al (2014) who undertook a 1.1 microsecond time length simulation of the MD of crystalline staphylococcal nuclease, providing 100-fold more sampling of time evolution than previous MD studies. Their aim was to compare the diffuse scattering from the simulated movements versus the experimental X-ray diffuse scattering measurements. They concluded that:

> *The total diffuse intensity calculated using the MD model is highly correlated with the experimental data ... (and that) diffuse scattering can be used to validate predictions from MD simulations and can provide information to improve MD models of protein motions.*

In the absence of diffuse X-ray scattering measurements how can the validity of an MD simulation be assessed? Pratap et al (2001) used multiple simulation runs with different starting conditions such as temperature. Thereby,

Precision and Accuracy in Biological Crystallography, Diffraction, Scattering, Microscopies, and Spectroscopies. John R. Helliwell, Oxford University Press. © John R. Helliwell (2025). DOI: 10.1093/9780198952848.003.0017

four sets of nanosecond time-length molecular dynamics (MD) simulations, two at 293 K and the other two at 313 K, were performed on each of two protein saccharide complexes. Pratap et al (2001) identified the simultaneous presence of direct lectin–carbohydrate interactions, e.g. 794 unique modes were found in 4000 snapshots in the four MD runs of the peanut lectin–T-antigen complex, one of the complexes studied. The corresponding number in their lactose complex was 540 unique modes. None of the unique modes were observed for more than 10% of the time, although there were close relationships amongst some of the modes (e.g. see Figure 17.1). Another cross check is to compare the RMS deviations of the atom positions in the MD ensemble with the RMS spread estimated from the crystallographic B factors (see section 4.4 of Bradbrook et al (1998) who found that these independent estimates were consistent).

Pratap et al (2001) cautioned that:

> *Detailed examination of the calculations points to the need for exercising caution in interpreting results of MD simulations: while long simulations are not possible owing to computational reasons, it is desirable (at least) to carry out several short simulations with somewhat different initial conditions.*

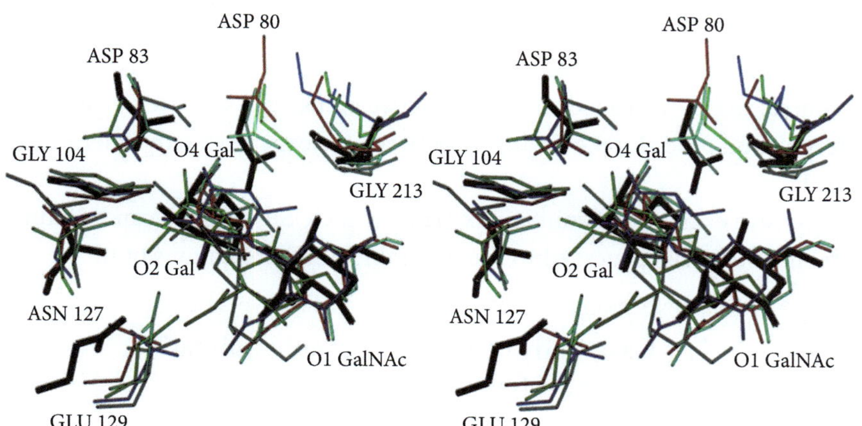

Figure 17.1 Stereoview of the locations of the representative modes I (red), II (green), III (cyan), IV (magenta), and V (grey) in the T-antigen complex. (These several colours are not important to visualize individually; the main point is that the precise structures of the five modes show a spread.) The location in the crystal structure is shown in black.

From Pratap et al (2001). This figure is reproduced with the permission of IUCr Journals.

The figure and text quote are reproduced with the permission of IUCr Journals. The feasibility of undertaking multiple simulations was made possible, needed at the time, by using a supercomputer which was installed at the Education and Research Centre of the Indian Institute of Science in Bangalore.

Recently Trewhella et al (2024) benchmarked molecular dynamics predictive methods for small-angle X-ray scattering from atomic coordinates of proteins using maximum likelihood consensus data.

Key Learning Points

- Starting from an experimental crystal structure, elucidating molecular dynamics of macromolecules is entirely feasible.
- Assessment of a molecular dynamics calculation can be made by carrying out several computations with different initial conditions.
- In special cases long timescale molecular dynamics runs can be assessed by comparison with diffuse scattering measurements.

17.2 Molecular animations

Animations are a powerful way towards an understanding of a molecular system's transition between two or more static structures. They are animations, of course, and the so-called 'morphing' between structures is based on a linear interpolation procedure to transition between the experimentally determined coordinates.

Extensive examples of animations made with UCSF Chimera can be found here: https://www.cgl.ucsf.edu/chimera/animations/animations.html

17.2.1 Case study of a β-crustacyanin animation and what we learnt from it

We calculated a molecular movie by linear interpolation based on two 'end-point' protein structures, i.e. apocrustacyanin A1 (Cianci et al (2001)) and A1 associated with the two astaxanthins in β-crustacyanin (Cianci et al (2002)), and this is described in Chayen et al (2003). This movie highlighted the structural changes forced upon the astaxanthin carotenoid on its complexation with the protein. In contrast, the protein-binding site remained relatively unchanged in the binding region, but there was a large conformational change occurring in a more remote surface-loop region, which, it was suggested, could be important in the complexation of astaxanthin and contribute to the spectral properties of the bound form in β-crustacyanin. The two endpoints of the animation are shown in Figure 17.2 and the movie can be activated by clicking on the gif provided in Chayen et al (2003).

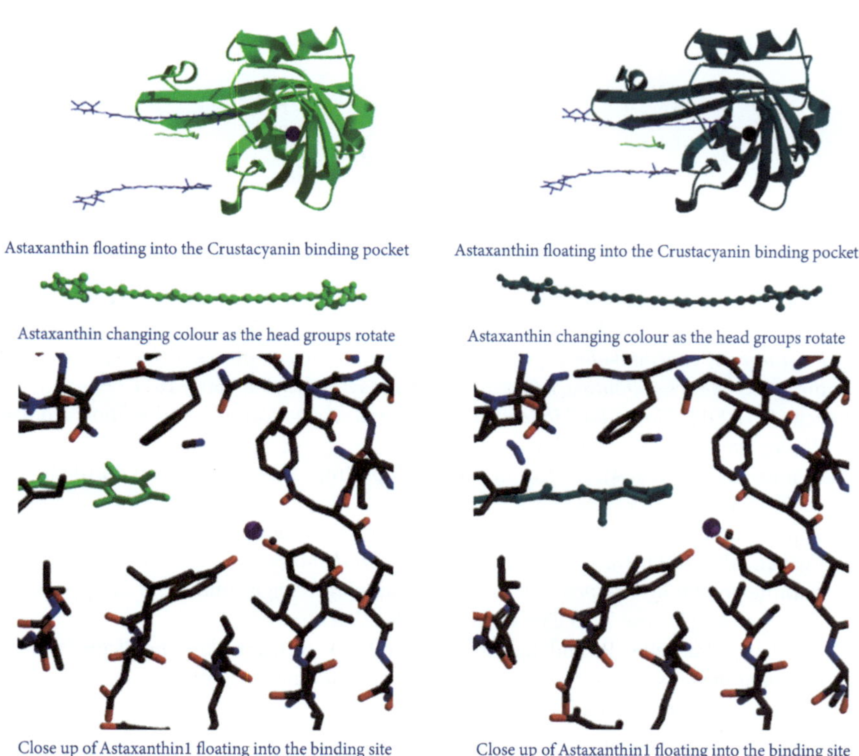

Figure 17.2 Two snapshots from the molecular animation, each with their own caption, based on two 'endpoint' protein structures, i.e. apocrustacyanin A1 (Cianci et al (2001)) and A1 associated with the two astaxanthins in β-crustacyanin (Cianci et al (2002) and Chayen et al (2003)). The animation was prepared by morphing (i.e. interpolating the coordinates of all the atoms between the two end points in 10 steps). An image was prepared for each step and those from steps 2 to 9 were repeated in reverse order to make steps 11 to 18. The full series was then compiled into a multipage image to make the final animation. The animation was developed by Dr P. J. Rizkallah, as described in this caption, our coauthor on these publications.

Key Learning Points

- Animations, tethered to experimental crystal structures, are like a cartoon. They can be very informative, albeit a guess of what happens between the frames of the cartoon.
- Animations are extensively used, and software tools are available to produce them such as ChimeraX (Pettersen et al (2004)).

References

References for section 17.1

Bradbrook, G. M., Gleichmann, T., Harrop, S. J., Habash, J., Raftery, J., Kalb (Gilboa), J., Yariv, J., Hillier, I. H., and Helliwell, J. R. (1998). *X-Ray and molecular dynamics studies of concanavalin-A glucoside and mannoside complexes: Relating structure to thermodynamics of binding.* J. Chem. Soc. Faraday Trans., **94**, 1603–1611.

Leach, A. (2001). *Molecular Modelling: Principles and Applications, 2nd Edition.* Addison Wesley Longman, Harlow, Essex, UK.

McCammon, J. A. (1984). *Protein dynamics.* Rep. Prog. Phys., 47(1). doi: 10.1088/0034-4885/47/1/001

Pratap, J. V., Bradbrook, G. M., Reddy, G. B., Surolia, A., Raftery, J., Helliwell, J. R., and Vijayan, M. (2001). *The combination of molecular dynamics with crystallography for elucidating protein ligand interactions: A case study involving peanut lectin complexes with T-antigen and lactose.* Acta Cryst., D57, 1584–1594.

Trewhella, J., Vachette, P., and Larsen, A. H. (2024). *Benchmarking predictive methods for small-angle X-ray scattering from atomic coordinates of proteins using maximum likelihood consensus data.* IUCrJ, 11, 762–779.

Wall, M. E., VanBenschoten, A. H., Sauter, N. K., Adams, P. D., Fraser, J. S., and Terwilliger, T. C. (2014). *Conformational dynamics of a crystalline protein from microsecond-scale molecular dynamics simulations and diffuse X-ray scattering.* Proc. Natl. Acad. Sci. U.S.A., 111(50), 17887–17892. https://doi.org/10.1073/pnas.1416744111

References for section 17.2

Chayen, N. E., Cianci, M., Grossmann, J. G., Habash, J., Helliwell, J. R., Nneji, G. A., Raftery, J., Rizkallah, P. J., and Zagalsky, P. F. (2003). *Unravelling the structural chemistry of the colouration mechanism in lobster shell.* Acta Cryst., D59, 2072–2082.

Cianci, M., Rizkallah, P. J., Olczak, A., Raftery, J., Chayen, N. E., Zagalsky, P. F., and Helliwell, J. R. (2001). *Structure of lobster apocrustacyanin A1 using softer X-rays.* Acta Cryst., D57, 1219–1229.

Cianci, M., Rizkallah, P. J., Olczak, A., Raftery, J., Chayen, N. E., Zagalsky, P. F., and Helliwell, J. R. (2002). *The molecular basis of the coloration mechanism in lobster shell: β-Crustacyanin at 3.2-Å resolution.* Proc. Natl Acad. Sci. USA, 99, 9795–9800.

Pettersen, E. F., Goddard, T. D., Huang, C. C., Couch, G. S., Greenblatt, D. M., Meng, E. C., and Ferrin, T. E. (2004). *UCSF Chimera—a visualization system for exploratory research and analysis (version 1.16).* J. Comput. Chem., 25, 1605–1612. Now this has transitioned to Chimera-X see https://www.cgl.ucsf.edu/chimerax/

18

Role of predictions as a grand challenge in biology

Protein fold prediction is solved

Protein folding prediction was one of the decades-old grand challenges of science. This had been a solely physics-based hunt for the Gibbs free energy minimum, which must be the protein 3D structure and identified by Anfinsen's (1973) protein denaturation and renaturation studies. However, it was not energy minimum methods that solved it; it was a combination of learning from protein sequences and precise protein experimental structures carefully curated in databases (Jumper et al (2021)). The continued wish for a physics-based, energy minimum approach has led to further debate on whether the protein folding challenge has been solved (Moore et al (2022)).

Protein design is a reverse procedure to protein fold prediction, which involves designing a protein structure and determining the amino acid sequence that would produce that designed protein. The first of these was published in 2003 and as remarked by the authors of Kuhlman et al (2003):

> *The ability to design a new protein fold makes possible the exploration of the large regions of the protein universe not yet observed in nature.*

They also recognized the potential for a sweeping change in protein crystallography phase determination; '*the design model was quite close to the true structure*' and could therefore be used for molecular replacement crystal structure determination.

On this latter point tuneable wavelength resonant scattering phasing have been somewhat supplanted by AlphaFold2; indeed, as Keegan et al (2024) found, with a large set of SAD-phased PDB depositions tested for solubility by molecular replacement using AlphaFold2 and other approaches, just 3% of cases were not solved and can remain then the domain of experimental-based phasing. In all cases any prediction of structure does not replace an experimental, refined, protein crystal structure. Beamlines with X-ray wavelength tuning

Precision and Accuracy in Biological Crystallography, Diffraction, Scattering, Microscopies, and Spectroscopies. John R. Helliwell, Oxford University Press. © John R. Helliwell (2025). DOI: 10.1093/9780198952848.003.0018

will still be required for those but likely to refocus their activities onto mainly site-specific metal or ion determination in macromolecular crystallography.

To come back to the themes of this book I note that both Kuhlman et al (2003) and Jumper et al (2021) stress, even in both of their article titles, that they have shown 'Design of a novel globular protein fold with atomic-level accuracy' and 'Highly accurate protein structure prediction'. Clearly, close-fit predictions to the current predominantly 100K database entries of precise protein X-ray crystal structures are an excellent achievement by both teams, worthy of the Nobel Prize for Chemistry 2024 to David Baker, Demis Hassabis and John Jumper (The Royal Swedish Academy of Sciences (2024)), but they are not accurate. Indeed, any protein 3D structure prediction or design must be experimentally validated by crystallography, cryoEM, or NMR spectroscopy. More than that though, the need is with us for an expanded number of experimental structures at a wide range of physiologically relevant conditions, ideally body temperature (Jacobs et al (2024)), if there are to be attempts at a close fit of predictions to the living cell situation and the molecular structure dynamics involved. As remarked earlier, X-ray radiation damage will be enhanced at this higher temperature and serial femtosecond or neutron crystallography may become necessary to side-step radiation damage.

Key Learning Points

- Protein fold prediction has been a grand challenge for many decades.
- A comprehensive solution has been found via the huge sequence and 3D experimental structure databases. A prediction is accompanied by a level of confidence in it.
- A physics-based, energy minimum solution to protein folding is still sought.
- A protein 3D structure prediction or design needs to be experimentally validated.
- Structural dynamics over very long timescales are not predictable at the present.

References

Anfinsen, C. B. (1973). *Principles that govern the folding of protein chains.* Science, 181, 223–230.

Jacobs, F. J. F., Helliwell, J. R., and Brink, A. (2024). *Body temperature protein X-ray crystallography at 37°C: A rhenium protein complex seeking a physiological condition structure.* Chem. Commun. https://doi.org/10.1039/D4CC04245J.

Jumper, J., Evans, R., Pritzel, A., Green, T., Figurnov, M., Ronneberger, O., Tunyasuvunakool, K., Bates, R., Žídek, A., Potapenko, A., Bridgland, A., Meyer, C., Kohl, S. A. A., Ballard, A. J., Cowie, A., Romera-Paredes, B., Nikolov, S., Jain, R., Adler, J., Back, T., Petersen, S., Reiman, D., Clancy, E., Zielinski, M., Steinegger, M., Pacholska, M., Berghammer, T., Bodenstein, S., Silver, D., Vinyals, O., Senior, A. W., Kavukcuoglu, K., Kohli, P., and Hassabis, D. (2021). *Highly accurate protein structure prediction with AlphaFold*. Nature, 596, 583–589.

Keegan, R. M., Simpkin, A. J., and Rigden, D. J. (2024). *The success rate of processed predicted models in molecular replacement: implications for experimental phasing in the AlphaFold era*. Acta Cryst., D80, 766–779.

Kuhlman, B., Dantas, G., Ireton, G. C., Varani, G., Stoddard, B. L., and Baker, D. (2003). *Design of a novel globular protein fold with atomic-level accuracy*. Science, 302, 1364–1368.

Moore, P. B., Hendrickson, W. A., Henderson, R., and Brunger, A. T. (2022). *The protein-folding problem: Not yet solved*. Science, 375, 507.

The Royal Swedish Academy of Sciences (2024). *The Nobel Prize in Chemistry 2024*, press release. https://www.nobelprize.org/uploads/2024/10/press-chemistryprize2024-2.pdf.

19

A new method

X-ray photon correlation spectroscopy (XPCS) to study biocondensed matter

XPCS is a method to measure dynamics on timescales of 1 s—10^{-12} s in condensed matter systems by correlating time-resolved coherent diffraction patterns to monitor nanoscale fluctuations such as in crowded protein solutions of a biological cell. Coherent X-rays, which are monochromatic, are required, as provided by an X-ray laser. The X-ray intensity is attenuated to avoid beam-induced effects. The scattering intensity, which fluctuates due to the changes of the speckle pattern, is recorded with a pixelated 2D array detector. The measurement of this fluctuating speckle pattern must balance reducing the random errors of the photon-counting fluctuations and containing the systematic error that would arise from radiation damage to the sample. This new method is under development and aims to provide time-resolved imaging of functionally relevant living processes (Perakis and Gutt (2020)).

The facilities particularly involved at present are the Stanford Linac Coherent Light Source (Alonso-Mori et al (2015)) and the EuroXFEL (Madsen et al (2021)). Applications to a wider range of examples will follow these early applications given the enthusiastic activity of the facilities and their user groups to develop this method.

On the name of the method, we have the three conditions of: a fixed wavelength monochromatic incident beam, an elastic scattering interaction and not a measured spectrum. So, this seems to me that XPCS should be called X-ray photon correlation scattering.

Precision and Accuracy in Biological Crystallography, Diffraction, Scattering, Microscopies, and Spectroscopies. John R. Helliwell, Oxford University Press. © John R. Helliwell (2025). DOI: 10.1093/9780198952848.003.0019

Key Learning Points

- This new method of X-ray Photon Correlation Spectroscopy (XPCS) utilises a coherent X-ray beam. It is under active development and has significant potential in domains of the study of the living cell that are otherwise difficult to measure.

References

Alonso-Mori, R., Caronna, C., Chollet, M., Curtis, R., Damiani, D. S., Defever, J., Feng, Y., Flath, D. L., Glownia, J. M., Lee, S., Lemke, H. T., Nelson, S., Bong, E., Sikorski, M., Song, S., Srinivasan, V., Stefanescu, D., Zhu, D., and Robert, A. (2015). *The X-ray Correlation Spectroscopy instrument at the Linac Coherent Light Source*. J. Synchrotron Rad., 22, 508–513.

Madsen, A., Hallmann, J., Ansaldi, G., Roth, T., Lu, W., Kim, C., Boesenberg, U., Zozulya, A., Moller, J., Shayduk, R., Scholz, M., Bartmann, A., Schmidt, A., Lobato, I., Sukharnikov, K., Reiser, M., Kazarian, K., and Petrov, I. (2021). *Materials Imaging and Dynamics (MID) instrument at the European X-ray Free-Electron Laser Facility*. J. Synchrotron Rad., 28, 637–649.

Perakis, F. and Gutt, G. (2020). *Towards molecular movies with X-ray photon correlation spectroscopy*. Phys. Chem. Chem. Phys., 22, 19443–19453.

20
Conclusions

What do I hope you have learned from this book? First and foremost, to be critical of what a single structure determination by each of the methods offers. This type of assessment is offered in the excellent book by Professor Sir Alan Fersht (1999) page 45. He answered the question: are the X-ray crystal and NMR solution structures of an enzyme essentially identical? His answer was that their correspondence was excellent albeit allowing for crystal lattice contact perturbations and the crystallization conditions required. He also highlighted that in many cases the crystal retained enzymatic activity. He also allowed for crystallization selecting a single conformation where there might be two or more.

Noting that firm endorsement, I would still press that without community agreed precision estimators of atomic coordinates or B factor values, how can two structures be declared identical? The chapters in this book will, I hope, provide skills for the researcher to consider more critically the results they look at, both the publications and their underpinning data.

More generally, education and training remain essential. A wide range of crystallographic societies worldwide work hard on providing these. One such recent example was organized by the IUCr at the European Crystallography Meeting in Vienna in 2019 (https://www.iucr.org/resources/data/commdat/vienna-workshop) and addressed data science skills for authors, editors, and referees. In the same year the first Crystallographic Information Fiesta (CIFiesta), an international crystallography school organized by the Commission on Teaching of the Italian Crystallographic Association (AIC), was celebrated in Naples (https://www.iucr.org/resources/cif/comcifs/cifiesta-2019). Despite all these efforts there are cases where post-publication peer review is needed and considerable work within that theme has been undertaken, perhaps most famously by Professor Dick Marsh with hundreds of corrected crystallographic space groups in the published literature (summarized by Ton Spek in a tribute lecture (http://www.cryst.chem.uu.nl/spek/ppp/marsh_toronto.pdf)). In structural biology notable examples involved the corrections to a number of covid 19 pandemic structures in the PDB (Jaskolski et al (2021) and Gao et al (2023)). Topical reviews of families of enzyme structures, including

Precision and Accuracy in Biological Crystallography, Diffraction, Scattering, Microscopies, and Spectroscopies.
John R. Helliwell, Oxford University Press. © John R. Helliwell (2025). DOI: 10.1093/9780198952848.003.0020

extensive tabulations of data entries, show up a wide range of incompleteness of deposition models (Helliwell (2021), Brink et al (2022), Hanau and Helliwell (2022, 2024)). Such topical reviews are a new category of article in *Acta Cryst F* (Helliwell et al (2021)) and can make an excellent starting point for the new PhD student to survey the landscape of biological structures within their thesis topic. Altogether these various efforts enhance the trust in our results, whilst charcterising the remaing uncertainties, which is as close as we can get to scientific truth (Helliwell (2024)).

Key Learning Points

- The question: 'are macromolecular structures from crystallography and from NMR the same?' is critically discussed. The absence of estimation of errors on coordinates and B factors hampers these discussions. This topic has run throughout the whole of this book.
- Recommendations are made for how newcomers to the field of structural biology can critically assess the models and data that they use in their research.
- More generally, education and training remain essential.
- Topical reviews of families of enzyme structures, including extensive tabulations of data entries, show up a wide range of incompleteness of deposition models.

References

Brink, A., Jacobs, F. J., and Helliwell, J. R. (2022). *Trends in coordination of rhenium organometallic complexes in the Protein Data Bank.* IUCrJ, 9 (2), 180–193.

Fersht, A. (1999) *Structure and Mechanism in Protein Science: A Guide to Enzyme Catalysis and Protein Folding*, 3rd Edition. W H Freeman, New York.

Gao, Y., Thorn, V., and Thorn, A. (2023). *Errors in structural biology are not the exception.* Acta Cryst., D79, 206–211.

Hanau, S. and Helliwell, J. R. (2022). *6-Phosphogluconate dehydrogenase and its crystal structures.* Acta Cryst., F78, 96–112.

Hanau, S. and Helliwell, J. R. (2024). *Glucose-6-phosphate dehydrogenase and its 3D structures from crystallography and electron cryo-microscopy.* Acta Cryst., F80, 236–251.

Helliwell, J. R., Newman, J., and van Raaij, M. J. (2021). *Topical reviews in Acta Crystallographica F Structural Biology Communications.* Acta Cryst., F77, 385.

Helliwell, J. R. (2021). *The crystal structures of the enzyme hydroxymethylbilane synthase, also known as porphobilinogen deaminase.* Acta Cryst., F77, 388–398.

Helliwell, J. R. (2024). *The Scientific Truth, the Whole Truth and Nothing but the Truth.* CRC Press, Boca Raton, Florida, USA.

Jaskolski, M., Dauter, Z., Shabalin, I. G., Gilski, M., Brzezinski, D., Kowiel, M., Rupp, B., and Wlodawer, A. (2021). *Crystallographic models of SARS-CoV-2 3CL^{pro}: In-depth assessment of structure quality and validation.* IUCrJ, 8, 238–256.

21

Appendix A1

Bayesian reasoning in data analysis and model refinement

Bayes' theorem was first derived by the eighteenth-century English clergyman and mathematician Thomas Bayes and found in his papers on his death. It was rediscovered by Laplace, who used it to solve a wide variety of outstanding problems in astronomy. I find the review article by Gilmore (1996) an excellent overview of Bayesian analyses in crystallography. [With the IUCr Journals permission I am guided by several sentences of Gilmore (1996) below.]

French (1978) advocated Bayesian methods of statistical analysis for crystallography, namely one which would provide a more powerful approach than classical least squares. The general principles of the method were discussed by French (1978) and applications to crystallographic refinement and primary reduction of diffractometer data were also described. For comprehensive details on Bayesian reasoning in data analysis see D'Agostini (2003).

In Chapter 2 I already drew attention to Schwarzenbach et al's (1989, 1995) excellent IUCr Subcommittee on Statistical Descriptors Reports which described the uncertainties of measurement as well as prior molecular knowledge. It is worth repeating their recommended overarching guideline of good practice that:

> ***Thoughtless use of established procedures in widely distributed software may be as harmful as the natural tendency of most people to prefer results in agreement with preconceived ideas.*** *Note, however, that preconceived ideas are an ingredient of Bayesian statistics. Since the precision may be evaluated with greater confidence than the accuracy, it is not surprising that the results of independent determinations of the same structure may differ by much more, and hardly ever by less than is allowed by statistical tests.*

The above text extract is reproduced with the permission of IUCr Journals.

Let's now explore Bayes' theorem, which at its simplest is as follows.

Let us suppose that we are carrying out an experiment to gain information on a parameter θ. Before commencing the experiment, we have prior belief

Precision and Accuracy in Biological Crystallography, Diffraction, Scattering, Microscopies, and Spectroscopies. John R. Helliwell, Oxford University Press. © John R. Helliwell (2025). DOI: 10.1093/9780198952848.003.0021

of the possible values of θ, which we express as an unconditional probability distribution p(θ). If the experiment yields data represented by the vector x, then Bayes's theorem states that:

$$p(\theta|x) \text{ is proportional to } p(x|\theta)\ p(\theta) \qquad \text{Equation A1.1}$$

The constant of proportionality may be determined by normalization. The term $p(x|\theta)$ is called the posterior distribution (or simply posterior); the function $p(x|\theta)$ is the likelihood, which takes account of the data from the experiment, and $p(\theta)$ is the prior probability.

As written, this is an uncontroversial equation simply derived from the axioms of probability theory and used by all statisticians to invert conditional probabilities, i.e. to obtain $p(x|\theta)$ from $p(\theta|x)$ and vice versa. However, Bayesian statisticians take the topic a whole step further by using Equation A1.1 as a model of scientific inference. In addition, probabilities are defined not in the narrow frequentist way (i.e. such as how often things occur) but in the broader, and much more controversial, sense of degree of belief, which is ideally suited to science where measurements can be very difficult to relate to frequencies of occurrence without the use of very convoluted argument, and a counter-intuitive use of language. Furthermore, Equation A1.1 is used iteratively: the posterior becomes an updated prior and the whole cyclic process repeated until convergence is reached.

Some everyday examples can be found in the book by Barnett (1972) (see his pages 11 and 12). In his terminology he uses H for a hypothesis and E is for evidence. Thereby, $P(H|E) = P(E|H)\,P(H)$ and $P(H|E)$ is then the probability of the hypothesis given the evidence. For example, he considers a drunk person flipping a coin. The person being drunk aspect is that they are overconfident and imagine that they can perfectly predict the heads versus tails outcomes of coin tosses. However, this is obviously a totally false belief in which one would have zero confidence. So, this can be expressed in Bayes' formalism that the drunk's hypothesis has a probability $P(H) = 0$. Therefore $P(H|E)$ will also be zero.

A very simple macromolecular crystallography example where $P(H) = 0$ would be diffraction data measured from a crystal where the unit cell size determined from the diffraction data is very small and so it can be deduced that the crystal was of a precipitating salt which had crystallized and not a macromolecule. Therefore, a model refinement, assuming a protein, would be pointless.

Improvements in structural biology methods described in this book, namely their enhanced throughput, open the door to a frequentist approach to using macromolecular structure model replicates. That said, even with atomic resolution X-ray diffraction data the more mobile parts of a macromolecular structure require the covalent bonding geometry restraints to be applied; an example comparing with and without restraints in a 0.94Å protein model refinement is that of Deacon et al (1997).

Key Learning Points

- Prior knowledge can be harnessed using Bayes' theorem and finds important application in molecular structure determination as restraints on covalent bond distances and angles.
- Without Bayes' theorem, methods such as macromolecular crystallography would routinely have insufficient experimental data to secure a good molecular 3D structure model.

References

Barnett, V. (1972). *Comparative Statistical Inference*, 2nd Edition. Wiley, New York.

D'Agostini, G. (2003). *Bayesian Reasoning in Data Analysis: A Critical Introduction.* World Scientific Publishing Company, Singapore.

Deacon, A., Gleichmann, T., Kalb (Gilboa), A. J., Price, H., Raftery, J., Bradbrook, G., Yariv, J., and Helliwell, J. R. (1997). *The structure of concanavalin A and its bound solvent deter mined with small-molecule accuracy at 0.94Å resolution.* Faraday Transactions, 93 (24), 4305–4312.

French, S. (1978). *A Bayesian three-stage model in crystallography.* Acta Cryst., A34, 728–738.

Gilmore, C. J. (1996). *Maximum entropy and Bayesian statistics in crystallography: A review of practical applications.* Acta Cryst., A52, 561–589.

Schwarzenbach, D., Abrahams, S. C., Flack, H. D., Gonschorek, W., Hahn, Th., Huml, K., Marsh, R. E., Prince, E., Robertson, B. E., Rollett, J. S., and Wilson, A. J. C. (1989). *Statistical descriptors in crystallography: Report of the IUCr Subcommittee on Statistical Descriptors.* Acta Cryst., A45, 63–75.

Schwarzenbach, D., Abrahams, S. C., Flack, H. D., Prince, E., and Wilson, A. J. C. (1995). *Statistical descriptions in crystallography. II. Report of a Working Group on Expression of Uncertainty in Measurement.* Acta Cryst., A51, 565–569.

Bibliography

List of related books and brief comments on how they differ from this book

Clearly my book's predominant theme is structural biology. To date we have books such as: Cantor and Schimmel's ***Biophysical Chemistry*** (1971), a three-volume opus whose assessment of a protein crystal structure is less than one page and focuses on the R factor (page 763). Overall, it received a favourable review here in Nature in 1981 https://www.nature.com/articles/289716b0.pdf.

Serdyuk, Zaccai, and Zaccai's (2007) ***Methods in Molecular Biophysics: Structure, Dynamics and Function*** (1120 pages) is a single-volume text which expands the assessment of a protein model beyond Cantor and Schimmel's 1971 books but not by much (pages 860 to 863 and 870 to 875). Within these few pages we have the trite, if not incorrect statement: '*The quality of a model depends on the quality of the data used in its creation*'. The need for precision estimators on atomic coordinates and B factors does not feature at all. The book is reviewed in Contemporary Physics, 59(3), 317–318 (2018), https://doi.org/10.1080/00107514.2018.1464514.

Daune's ***Biophysique Moleculaire*** (1993), translated into English and published by OUP in 1999, is endorsed by David Blow FRS in his Foreword but within which he states in an almost throwing up his hands on assessing the precision and accuracy of biological systems '*Daune's treatment is rigorously built on basic physical principles and a mathematical analysis. Each concept is developed into the dreadful complications of real biological systems. The role of experimental data is to demonstrate that such development is soundly based, or to show where further analysis is required.*' Actually, a rather insightful resumé of '*The significance of the B factor*' is provided on page 218. The book is reviewed here: https://www.ncbi.nlm.nih.gov/pmc/articles/PMC1225493/pdf/biophysj00071-0417.pdf.

As Serdyuk et al's book explicitly recognizes in its title (***Methods in Molecular Biophysics: Structure, Dynamics and Function)***, if one wants to understand function based on structure (i.e. the famous 'structure–function relationship'), then dynamics must be explicitly treated. The B factor (i.e. the ADP, atomic displacement parameter, of chemical crystallography) is one approach and the other is time-resolved methods. Hence the recent book ***Dynamics and Kinetics in Structural Biology: Unravelling Function Through Time-Resolved Structural Analysis*** by Moffat and Lattman (2023) is a significant arrival on the scene. I reviewed this book in IUCr Journals https://journals.iucr.org/d/issues/2024/03/00/xo0199/index.html. The growth of the field of structural dynamics must surely ask the question: how can we decide or agree that an atom has moved significantly or not if there are no coordinate error descriptors on a structure? It received no treatment in Moffat and Lattman's book as I remarked as follows '*The archiving of raw diffraction data is very important as these rapidly changing data analytics can then be applied to the actual experimental measurements and assess the reproducibility of earlier analyses and interpretations in publications. This is especially so where diffraction data changes are small, the occupancy of initiated intermediates are low and especially where there is an absence of error estimation of atomic coordinates or B factors (Helliwell (2023), Current Research in Structural Biology*

6, 100111). However, it is good that the role of raw diffraction data is acknowledged on page 163'.

The famous teaching of crystallography book by Glusker and Trueblood, ***Crystal Structure Analysis: a Primer,*** has, right from the 1st edition (1972), a very good three-page resumé of precision and accuracy aspects in chemical crystallography (pages 118 to 121) in a section neatly entitled '*The correctness of a structure*'.

For other areas of structural science in biology the book by Joachim Frank published in 2006 is an important one: ***'Three-Dimensional Electron Microscopy of Macromolecular Assemblies: Visualization of Biological Molecules in Their Native State'.*** The section entitled '*Validation and Consistency*' (pages 286 to 291) is a useful overview of good sense assessments of cryoEM structures and offers a useful starting point for my book's content on the precision and accuracy of cryoEM.

A comprehensive treatment of the various biological structure techniques are the chapters in *International Tables Volume F, 2nd edition* (2011), Part 19, pages 563 to 632. Likewise, a range of overviews of structure validation are in Part 21. Section 18.5 is a chapter by Cruickshank himself on 'Coordinate uncertainty'.

Abbreviations List

3DEM	3-dimensional electron microscopy
ACA	American Crystallographic Association
BNL	Brookhaven National Laboratory, USA
CCDC	Cambridge Crystallographic Data Centre
CCP4	Collaborative Computational Project Number 4 in Protein Crystallography
CCPBiosim	Collaborative Computational Project in Biomolecular Simulation
CCP-EM	Collaborative Computational Project for Electron cryo-Microscopy
CIF	Crystallographic Information Framework
ECA	European Crystallographic Association
EPR	Electron Paramagnetic Resonance
ESRF	European Synchrotron Radiation Facility
ESRP	European Synchrotron Radiation Project
ESI MS	Electrospray ionization mass spectrometry
EXAFS	Extended X-ray Absorption Fine Structure
ILL	Institut Laue Langevin
IUCr	International Union of Crystallography
NIST	National Institute of Standards and Technology, USA
NMR	Nuclear Magnetic Resonance
PDB	Protein Data Bank
SANS	Small-Angle Neutron Scattering
SAXS	Small-Angle X-ray Scattering

Index

For the benefit of digital users, indexed terms that span two pages (e.g., 52–53) may, on occasion, appear on only one of those pages.

Tables, figures, and boxes are indicated by an italic *t*, *f*, or *b*.